Fifth Edition
Laboratory Anatomy of the
Shark

Laurence M. Ashley
Formerly Professor of Zoology,
Walla Walla College,
College Place, Washington

Robert B. Chiasson
Professor of Veterinary Science
University of Arizona
Tucson, Arizona

wcb
Wm. C. Brown Publishers
Dubuque, Iowa

Laboratory Anatomy Series

Consulting Editor
  Robert B. Chiasson

Laboratory Anatomy of the Cat
  Ernest S. Booth
  Robert B. Chiasson

Laboratory Anatomy of the Fetal Pig
  Theron O. Odlaug

Laboratory Anatomy of the Frog
  Raymond A. Underhill

Laboratory Anatomy of the Human Body
  Bernard B. Butterworth

Laboratory Anatomy of the Mink
  David Klingener

Laboratory Anatomy of the Perch
  Robert B. Chiasson

Laboratory Anatomy of the Pigeon
  Robert B. Chiasson

Laboratory Anatomy of the Rabbit
  Charles A. McLaughlin
  Robert B. Chiasson

Laboratory Anatomy of the Shark
  Laurence M. Ashley
  Robert B. Chiasson

Laboratory Anatomy of the Turtle
  Laurence M. Ashley

Laboratory Anatomy of the White Rat
  Robert B. Chiasson

Illustrated by Robert B. Chiasson

Copyright © 1950 by Laurence M. Ashley

Copyright © 1969, 1976, 1983, 1988 by Wm. C. Brown Company Publishers. All rights reserved.

ISBN 0-697-05121-8

No part of this publication may be reproduced, stored in a retrieval system, or transmitted, in any form or by any means, electronic, mechanical, photocopying, recording, or otherwise, without the prior written permission of the publisher.

Printed in the United States of America
10 9 8 7 6

Contents

  List of Figures **iii**
  Preface **v**
  Introduction **vii**
1 External Anatomy and Skin, **1**
2 Skeletal System, **9**
3 Muscular System, **16**
4 Membranes, Mesenteries, and Coelomic Cavities, **28**
5 Oral and Pharyngeal Cavities, **32**
6 Digestive System, **36**
7 Circulatory System, **40**
8 Excretory System, **54**
9 Reproductive Systems, **58**
10 Nervous System, **61**
11 Special Sense Endings **71**
  Index, **79**

List of Figures

1. Phylogenetic chart of the vertebrates. **viii**
2. Anatomical planes and directions of the shark body. **xiv**
3. Dorsal and cross section view of the shark illustrating the relationship of body form to swimming. **1**
4. Lateral and frontal views of the shark illustrating "pitch" and "roll." **2**
5. Different types of tails found in fishes. **3**
6. Lateral view of a male dogfish shark. **4**
7. Lateral view of the shark head illustrating external characteristics. **4**
8. Ventral views of the pelvic regions of the shark. **6**
9. Photomicrograph of the placoid scales of the dogfish shark × 500. **6**
10. Section of the skin with a placoid scale. **7**
11. Section through the skin of the shark showing successive stages in the development of a scale. **8**
12. Dorsal and ventral views of the chondrocranium. **9**
13. Lateral view of the cranium, jaws, branchial basket, and pectoral girdle of the shark. **11**
14. Ventral view of the cranium, jaws, and branchial basket of the shark. **11**
15. Diagrammatic lateral view of the entire shark skeleton. **13**
16. Oblique views of trunk vertebrae, entire, sectioned, and frontal views of a caudal vertebra. **14**
17. Ventral views of the left halves of the pectoral (left) and pelvic (right) girdles and limbs of the shark. **15**
18. Diagram of dissection incisions. **17**
19. Oblique view of a cross section of the tail of the shark illustrating the relationships of muscle bundles to myotomes. **18**
20. Lateral view of the musculature of the shark and detail of a single myotome. **18**
21. Diagram of myomere movement. **19**
22. Lateral view of the head and branchial muscles. **20**
23. Ventral view of the superficial jaw and branchial muscles. **21**
24. Ventral view of the deep jaw muscles. **23**
25. Diagrammatic lateral view of the deep branchial muscles. **23**
26. Ventral view of the pelvic musculature of a female shark. **26**
27. Diagram of dissection incisions. **28**
28. Diagrams of vertebrate serous (coelomic) membrane formation. **29**
29. Ventral view of the shark visceral cavity and cross section diagrams at three different levels of the body cavity to illustrate the coelomic membranes. **31**
30. Cross section through the lower jaw (Meckel's cartilage) of a shark (*Negaprion brevisostris*) with the superficial tissue removed. **33**
31. Surface features of an opened gill arch and a gill arch with the filaments removed to show the gill rays and interbranchial muscles. **34**
32. Ventral view of the shark viscera and heart. **37**
33. Ventral view of the digestive tract dissected open to expose the mucosal surface of the tract and views of light microscope sections of the mucosa of the stomach and typhlosole. **38**
34. Midsagittal section of the shark heart *in situ* including attachments of the pericardial membrane. **41**
35. Diagrammatic sagittal views of the shark heart. **42**
36. Ventral view of the heart and aortic arches of the shark. **43**

37 Major blood vessels of the gill arches and cross section of a gill arch. **44**

38 Dorsal view of the heart and ventral aorta and ventral view of the major vessels of the branchial arches. **45**

39 Distribution of the internal carotid on the ventral surface of the brain. **46**

40 Ventral view of the aorta and lateral abdominal veins of a male shark. **48**

41 Ventral view of the body cavity with the viscera removed to expose the dorsal aorta and its major branches. **49**

42 Diagrammatic ventral view of the cardinal venous system of the shark. **50**

43 Ventral diagrammatic view of the cardinal venous system. **51**

44 Ventral view of the body cavity of the male shark illustrating the urogenital system. **55**

45 Diagrammatic reconstruction of a kidney tubule of the shark and its blood supply. **56**

46 Ventral diagrammatic views of the pregnant female shark and nonpregnant female genital system. **58**

47 Diagram of the shark brain and its divisions. **62**

48 Ventral view of the brain and cranial nerves. **63**

49 Dorsal view of the brain. **66**

50 Lateral view of the brain. **68**

51 Sagittal section of the shark brain. **69**

52 Cross section of the spinal cord and vertebra to illustrate the arrangement of the spinal nerves and gray and white matter of the spinal cord. **70**

53 Lateral view of the shark head illustrating external characteristics. **71**

54 A neuromast and hair cells (*A* and *B*) of the lateral line canal system. **72**

55 Dorsal view of the ears of the shark through a transparent chondrocranium. **73**

56 Lateral view of the ear of the shark through a transparent chondrocranium. **73**

57 Transverse section through the olfactory sacs and rostrum of the shark showing the interior of the olfactory sac and the ampulla of Lorenzini. **75**

58 Dorsal and ventral views of the shark eyeball and ocular muscles. **75**

59 Sagittal section of the shark eye. **77**

Preface

The major objectives of this revision were to correct errors and to make explanations clearer than in the past editions. To accomplish this, I have had the help of very critical and constructive reviewers who have pointed out ambiguities, inconsistencies, omissions, and other errors. I, therefore, would like to acknowledge the following reviewers for their valuable assistance: Dean G. Dillery, Albion College; Paul Fell, Connecticut College; Joseph Frantz, Los Angeles Valley College; James Hanken, University of Colorado–Boulder; Robert J. Raikow, University of Pittsburgh.

The evolution of this laboratory manual has occurred over the past 37 years. Dr. Ashley's first edition was one of the earliest guides to be well illustrated. Other dissection guides were dependent on written descriptions to provide instructions for the student, and the student was expected to provide his own illustrations (which were often part of the course grade). Then, as now, the primary (if not the exclusive) use of a shark dissection guide was in Comparative Vertebrate Anatomy courses. In 1950 Comparative Anatomy was a required prerequisite for premedical students and these courses had high enrollments. In only a decade nearly all medical schools dropped the requirement, and in the 1970s the course was only a recommendation in a few schools. The subsequent decline in enrollment in Comparative Anatomy courses has had a mixed reaction from course instructors. Some instructors argue that less quantity may improve the quality but textbooks and dissection guides decline in sales. Nevertheless this manual has survived, now appears in its fifth edition, and we are deeply gratified by its success.

Introduction

Depending upon what is accepted as shark characteristics, sharks probably originated in the Devonian period approximately 400 to 450 million years ago (fig. 1). The first sharks were members of the order Cladoselachii. Characters that identify these sharks concern the jaws, the way the jaws are jointed with each other and with the skull, and the shape and arrangements of the teeth. Other characters involve the fins and vertebral column.

Professor E. A. Stensiö, over 20 years ago argued that the Cladoselachii are derived from the most primitive jawed fishes, the Placodermi. Professor Stensiö further argued that the Placoderms are closely related to the Devonian sharks and therefore both should be included in the same major grouping. His argument is now fairly well accepted and although classifications differ from one author to the next, they all incorporate the Placodermi and Chondrichthyes (cartilaginous fishes), in one way or another, in the same group.

The following classification is taken from J. S. Nelson (1976) *Fishes of the World,* John Wiley & Sons, New York. The lowest grouping is to order except for Squaliformes, which is categorized to subfamily.

GRADE Pisces
  SUBGRADE Elasmobranchiomorphi
    CLASS Placodermi
      Order Arthrodiriformes
      Order Ptyctodontiformes
      Order Phyllolepiformes
      Order Petalichthyiformes
      Order Rhenaniformes
      Order Antiarchiformes
    CLASS Chondrichthyes
      SUBCLASS Elasmobranchii
        Superorder Cladoselachimorpha
          Order Cladoselachiformes
          Order Cladodontiformes
        Superorder Xenacanthimorpha
          Order Xenacanthiformes
        Superorder Selachimorpha
          Order Ctenacanthiformes
          Order Hybodontiformes
          Order Heterodontiformes
          Order Hexanchiformes
          Order Lamniformes

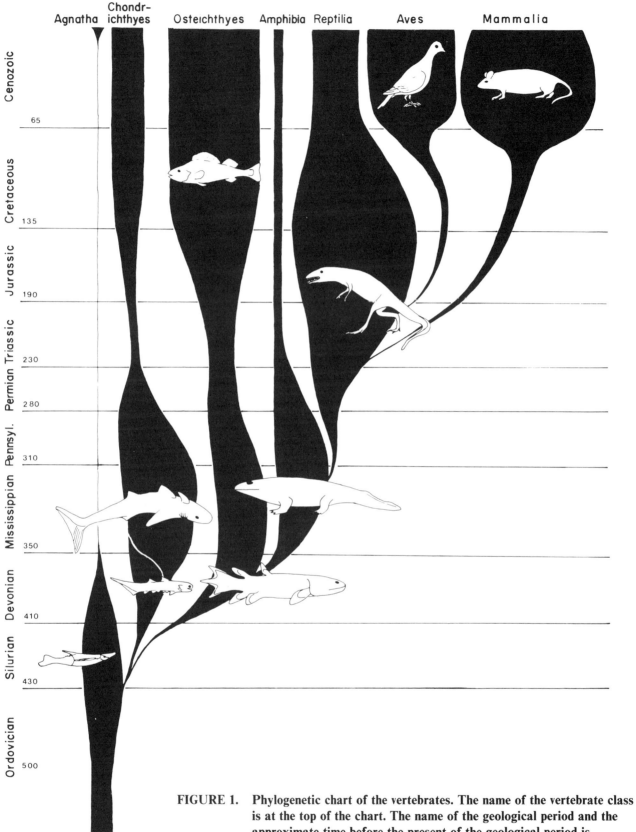

FIGURE 1. Phylogenetic chart of the vertebrates. The name of the vertebrate class is at the top of the chart. The name of the geological period and the approximate time before the present of the geological period is presented on the left of the chart. Time is presented in millions of years. The class Placoderma is unlabeled but is represented as a segment of the class Chondrichthyes. Ostracoderms are included with the Agnatha and the Acanthodia are included with the Osteichthyes.

Order Squaliformes
　Suborder Squaloidei
　　Family Squalidae
　　　Subfamily Dalatiinae
　　　Subfamily Echinorhininae
　　　Subfamily Squalinae (Dogfish Sharks)
　　　　Eight genera with 44 species including *Squalus acanthias*
　　Family Pristiophoridae
　　Family Squatinidae

A few authors do not accept the ordinal ending "-iformes" for fish and thus list *Squalus* as belonging to the order Selachii. This designation occurs in some recent textbooks of Comparative Anatomy that list only four orders of Elasmobranchii: Cladoselachii, Pleurocanthodii, Selachii, and Batoidea.

Just as the subgrade Elasmobranchiomorphi places the Placoderms closer to the Chondrichthyes than to the Osteichthyes, or bony fish, the subgrade Teleostomi includes the class Acanthodii with the Osteichthyes. These relationships are somewhat speculative and consequently it appears better to use the unofficial category of subgrade than the more official term "class" for these groups.

On the other hand those authors that believe more strongly in these relationships refer to the Elasmobranchiomorphi and Teleostomi as classes and Placodermi and Chondrichthyes as subclasses. This arrangement is also found in Comparative Anatomy textbooks.

Following Nelson's groupings the spiny dogfish shark is classified as follows:

Phylum Chordata
　Subphylum Vertebrata
　　Superclass Gnathostomata
　　　Grade Pisces
　　　　Subgrade Elasmobranchiomorphi
　　　　　Class Chondrichthyes
　　　　　　Subclass Elasmobranchii
　　　　　　　Superorder Cladoselachimorpha
　　　　　　　　Order Squaliformes
　　　　　　　　　Suborder Squaloidei
　　　　　　　　　　Family Squalidae
　　　　　　　　　　　Subfamily Squalinae
　　　　　　　　　　　　Tribe Squalinae
　　　　　　　　　　　　　Genus *Squalus*
　　　　　　Species *S. acanthias*
Scientific Name: *Squalus acanthias*.

Figure 1 diagrammatically illustrates the occurrence in time and numbers of species of the major groups of vertebrates. This figure should also help to place the shark in perspective in regard to its relationships to other vertebrates.

Thus considering availability as well as primitive and representative structures, the shark is a good compromise representative of all fishes.

HINTS FOR DISSECTING

Scissors or scalpel may be used when cutting through skin or cartilage but the dissection or separation of muscles is best accomplished with blunt dissection, that is, with the use of a blunt probe or the blunt edge of closed scissors. A pair of closed hemostatic forceps may also be useful for separating loose connective tissue from attached organs.

The following list of instruments may be purchased in a kit or individually and in various qualities and prices. Instruments of good quality available from a biological supply company catalog at the time of publication of this manual totaled $20. They should cost less at the school bookstore.

Scalpel handle #3
Scalpel blades #11 (package of 6)
Hemostatic forceps (Kelly or Mosquito type)
Forceps, medium, blunt tips
Scissors, surgical grade, one blunt tip
Mall probe, blunt tip
Dissecting needles, wood or plastic handle
Surgical gloves, nonsterile, box of 100

When dissecting:

1. Know the association of other structures to the one you plan to cut before you cut—avoid deep cuts.
2. Learn the anatomical terms that apply to the shark and use these terms when asking questions. Careful students will resolve most problems by themselves. If after earnest effort you cannot find a structure, then ask your instructor for help.
3. Compare your dissection specimen with that of your classmates. In this way you will see variations and features you might otherwise have overlooked. Do your own dissection first—then look around.

PLANES AND DIRECTIONS

The planes and directions on the shark body are illustrated in figure 2. The terms used here are adapted from several sources. There is no established set of terms for fishes (or for amphibians or reptiles) so no terms applied to the shark can be called incorrect, but we have avoided the use of *anterior* and *posterior* for the same reasons that these terms are avoided by avian and mammalian (other

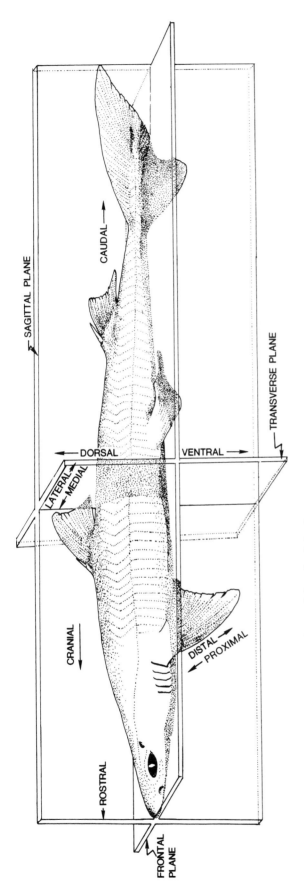

FIGURE 2. Anatomical planes and directions of the shark body.

than human) anatomists. That is, because the anatomical position of the human is erect, anterior of the human is ventral in horizontal vertebrates and consequently is a confused terminology.

Cranial (toward the head) and *rostral* (toward the snout; within the head) are used in place of anterior, and *caudal* (toward the tail) is used rather than posterior.

The following terms are also illustrated in figure 2.

Dorsal—toward the back or top of the shark.
Frontal plane—an imaginary plane separating the shark into dorsal and ventral sections.
Lateral—toward the side.
Medial—toward the midline.
Sagittal plane—an imaginary plane separating the shark into right and left halves.
Transverse plane—an imaginary plane separating the shark into rostral or cranial and caudal portions. This may be at any level from rostral to caudal.
Ventral—toward the bottom or belly of the shark.

SUGGESTED READINGS

Compagno, L. J. V. 1977. Phyletic relationships of living sharks and rays. *Amer. Zool.* 17 (2): 303–22.

Daniel, J. F. 1934. *The elasmobranch fishes.* 3d ed. Berkeley: University of California Press.

Nelson, J. S. 1976. *Fishes of the world.* New York: John Wiley & Sons.

Romer, A. S. 1966. *Vertebrate paleontology.* 3d ed. Chicago: University of Chicago Press.

Stensiö, E. A. 1963. Anatomical studies on the arthrodiran head. Part 1. Preface, geological and geographical distribution, the organization of the arthrodires, the anatomy of the head in the Dolichothoraci, Coccosteomorphi and Pachyosteomorphi. Taxonomic Appendix. *Handl. K. Svenska Vetenskapsakad.* 9 (2): 1–419.

Webb, P. W., and D. Weihs, eds. 1983. *Fish biomechanics.* New York: Preager Publishers.

Zangerl, R. 1973. Interrelationships of early chondrichthyians. In P. H. Greenwood, R. S. Miles, and C. Patterson, eds. Interrelationships of fishes, supp. 1. *Zool. J. Linnean Soc.* 53: 1–14.

Chapter 1
External Anatomy and Skin

EXTERNAL ANATOMY

The most obvious features of the shark's appearance are: (1) the streamlined shape of the body and (2) the fins (paired and unpaired) that break the otherwise smooth contours of the shark's body. These features are adaptations for swimming and provide the animal with advantages over its prey that have contributed to the long phylogenetic success of sharks.

Body Shape

Figure 3 is a dorsal view of a swimming shark and representative cross sections (*A* through *D*) illustrating the cranial-caudal changes in body shape. The snout and head are compressed dorsoventrally (*A*), but this changes in the area of the pharynx (gill slit region) to a roughly triangular shape (*B*), and is even more triangular at the level of the first dorsal spine and fin (*C*). Caudally, the shape changes once again to a round cross section (*D*).

There are several different categories for swimming by fishes based on body motions (or lack of) and the use (or nonuse) of the caudal fin. Other fish may use the paired fins in "rowing" movements, and still others may use the dorsal (the "Bowfin," *Amia*) or ventral fin (*Gymnotus,* an electric fish without a caudal fin) for propulsion. Most fish use the body and caudal fin with sinuous, eellike body motions (Anguilliform) or a more limited body "wave" movement (Carangiform or Subcarangiform). The swimming form of small sharks is *anguilliform* (named for the eel, *Anguilla*), which may be described as sinuous contortions of the body. In *Squalus acanthias* the entire body is thrown into at least a half-wavelength with a usually large amplitude over the length of the body. As the body bends in its wave formation, a thrust is produced against the water and it is this thrust that propels the animal forward assisted by movements of the heterocercal tail (see fig. 3 and discussion of caudal fin, pp. 2–4).

The Fins

Two prominent midline dorsal fins, each bearing a strong sharp spine, aid in protecting the shark from enemies. The spines carry a poison into the flesh of a man or animal wounded by them. The poison is expressed from soft, glandular tissue around the base of the spine and passes along a shallow groove on the caudal surface of the spine to its pointed end and into the wound. Such wounds are said to be quite painful.

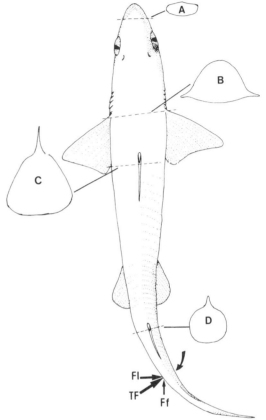

FIGURE 3. Dorsal and cross section views of the shark illustrating the relationship of body form to swimming. The propulsive force is generated by lateral movements of the tail and body, which are opposed by the resistance of the water (*TF*) producing a force in opposition to the body movement (curved arrow). This total force (*TF*) may be divided into two vectors representing a lateral component (*Fl*) and a longitudinal component (*Ff*). The propulsive longitudinal component is opposed by frictional drag, which is greatly reduced by the *streamlining* contour of the body.

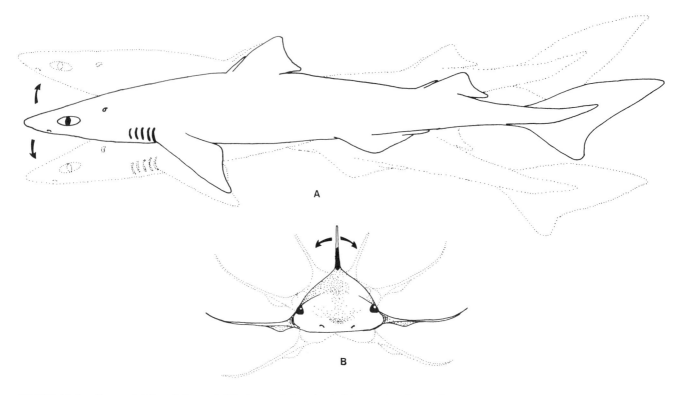

FIGURE 4. Lateral (A) and frontal (B) views of the shark illustrating "pitch" and "roll." Pitch is the "teeter-totter" effect that usually tilts the head down as the heterocercal tail propels the shark forward. Lowering the pectoral fins compensates for forward pitch. Roll rotates the body around its central axis and is prevented by the pectoral and dorsal fins.

In addition to the unpaired fins, sharks also have ventrally placed paired fins. These are the *pectoral* fins, located just caudal to the gills, and the *pelvic* fins, placed midway between the pectoral and caudal fins, and on either side of the cloacal aperture.

The Body Fins

In squaloid sharks such as *Squalus,* there is no anal fin and the pectoral fins are positioned higher on the side of the body than in other sharks. This placement of the pectoral fins puts them in line with the center of balance, the center of thrust from the dorsal heterocercal tail lobe, and with the flat under surface of the head.

Working in concert, the tail and pectoral fins turn the shark in nearly any direction. The flat head may also act as a fin, especially in vertical turns, which are important in feeding actions.

The dorsal and pelvic fins serve primarily as stabilizers preventing roll and pitch (fig. 4).

Sharks have no swim bladder for buoyancy as do most teleost (bony) fish. Therefore as soon as a shark stops swimming it begins to sink. Fish with homocercal (equal-lobed) tails and swim bladders can rest at various depths without sinking because gases passed into the swim bladder provide an adjustable buoyancy mechanism (hydrostatic organ) for flotation.

The Caudal Fin

The unpaired *caudal* (tail) fin (fig. 5) has a large dorsal vane and a much smaller ventral vane. Such a fin is termed *heterocercal* as contrasted with the more symmetrical *homocercal* caudal fins of teleost (bony) fishes. With the exception of those fish that have the terminal caudal muscles modified as electric organs (*Raja,* the electric skate; *Gymnotus,* the electric "eel"; and others), all fish have a caudal fin that is important as the major propulsive swimming organ. When the tail is moved laterally it creates a thrust against the water (fig. 3). The body is pushed in the opposite direction of the thrust of the tail in what we will call the *force* (caused by the inertia of the water and having a forward component, Ff, and a lateral component, Fl). Lateral movement is resisted by the water and tends to cause the fish body to rotate around its center of mass, but this contralateral movement of the head is also resisted by the water so the side-to-side movements are not great. The forward component has less resistance due to the streamlined shape of the fish and the fish is propelled forward (fig. 3). The

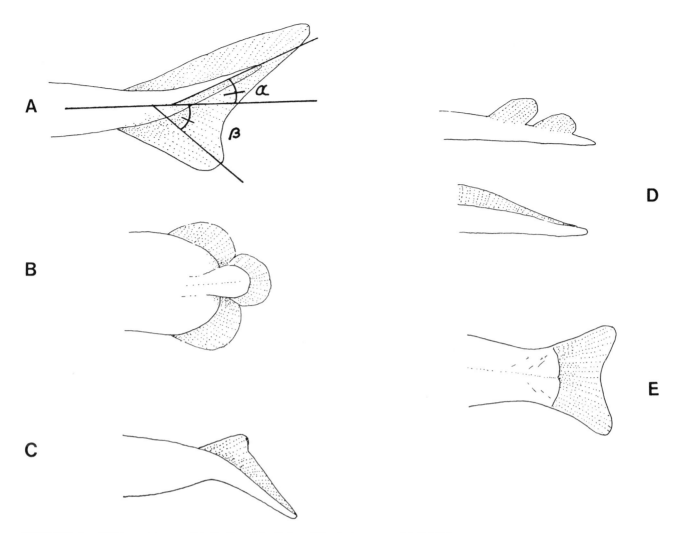

FIGURE 5. Different types of tails found in fishes. The heterocercal tail (A) is typical of sharks and is correlated with a horizontal arrangement of the pectoral fins (see text, pp. 2–4) Crossopterygian fishes (including the ancestor of tetrapods) have a diphycercal tail (B). The extinct anapsids (see fig. 1) had a hypocercal tail (C), and most modern bony fishes have a homocercal tail (E). Electric fishes such as the elasmobranch, *Raja* (Rayfish) or the teleost, *Gymnarchus* (electric eel) have tail musculature modified for producing an electrical charge and these muscles are not functional in locomotion. Consequently the tails have no caudal fins (D). Heterocercal angle = α; hypocercal angle = β.

movements of the tail are actually very complicated and produce forces in several directions in addition to those mentioned above. Friction and drag tend to work against forward movement.

The Heterocercal Tail

The heterocercal (*hetero-* = different; *cercal* = tail) tail has two lobes of different sizes and directed at different angles from the long axis of the body. The larger, heterocercal, lobe is directed upward and the smaller, hypochordal, lobe is directed downward.

The *heterocercal* angle is the acute angle formed by the lines of the long axis of the body and the line from the base of the tail to the point of the dorsal caudal fin lobe (fig. 5A). The *hypocercal* angle is between the same body axis line and a line from the base of the tail to the point of the ventral caudal fin lobe (fig. 5A). The terminal end of the tail projects into and supports the dorsal lobe. Each lateral tail movement produces two component forces, forward and transverse. The forward component is also directed downward by the heterocercal tail. The transverse component produces the only upward force. Both the

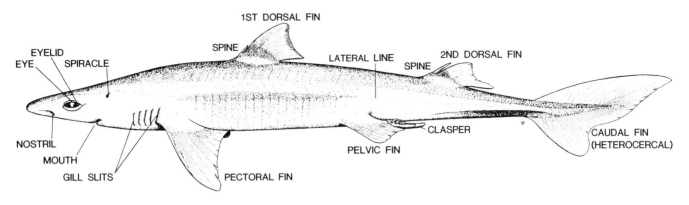

FIGURE 6. Lateral view of a male dogfish shark.

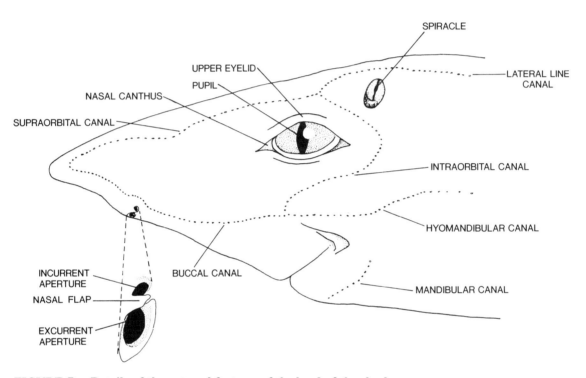

FIGURE 7. Details of the external features of the head of the shark.

degree of upsweep of the tail, that is the heterocercal angle, and the degree of rotation of the tail influence the forward and transverse forces that are produced.

The ratio of forward to transverse force changes as the speed of the tail beat and the angle of upsweep of the tail changes. The forward force is always greater than the transverse force so the upward force produced by the transverse component is always less than the downward force. Considering only the dorsal lobe of the shark's tail and by calculating the angle of the upsweep of the tail and the forces exerted by the tail movements, it has been shown that the optimal line of forward thrust from the tail passes through the shark's center of balance (fig. 4A). This center of balance might be likened to the fulcrum on a "teeter-totter" so the "teeter-totter" remains balanced as long as the "line of thrust" passes through the fulcrum but if the line of thrust passes behind the fulcrum (i.e., between the source of the force and the fulcrum) the opposite end is tilted upward. It has been calculated that tail angles of less than 33 degrees produces a "balanced thrust." If the heterocercal tail angle is greater than 33 degrees, the line of thrust falls behind the center of balance and the shark's head is tilted upward.

The ventral lobe of the shark's tail (the hypochordal lobe) has a ventral, hypochordal angle producing thrusts that are the opposite of the dorsal heterocercal lobe. If the line of thrust from the hypochordal lobe falls behind the center of balance, the head of the shark is directed downward. Thus the hypochordal lobe modifies the dorsal heterocercal lobe. Those sharks with a very abrupt heterocercal tail angle also have a strong hypocercal lobe. The dogfish shark has a heterocercal angle less than 33 degrees and a moderately developed hypochordal lobe. The hypochordal lobe may be absent in sharks with a very low heterocercal tail angle.

The Head

Find all of the structures labeled in figures 6 and 7 on your shark and learn their names and characteristics. The *head* extends from in front of the pectoral fins to the tip of the *rostrum* (snout), the *trunk* extends from the front of the pectoral fins to the pelvic fins, and the *tail* region is from the pelvic fins to the tip of the caudal fin. These three body regions are useful in anatomical descriptions.

The pale *lateral line* appears as a series of minute openings along both sides of the body, with connecting branches below and above the eye, and on the under side of the snout (figs. 7 and 53; also see pp. 71–72). Beneath the surface of this line is the *lateral line canal*, a special sensory organ to be described later as the *lateralis* system (chapter 11).

The head is dorso-ventrally compressed (fig. 3) and includes a forward projecting snout. Figure 7 illustrates the major features of the head.

1. The *rostrum* (snout) projects forward from the gills, mouth and eyes to form a tapered wedge or "prow."
2. Beneath the snout on either side are the nostrils (nares). Each *naris* consists of a rostral *incurrent aperture* and a more caudal *excurrent aperture*. The two apertures are separated externally by a transverse *nasal flap* (fig. 7B). Water enters the incurrent aperture and leaves via the excurrent aperture, thus permitting awareness of changing odors in the shark's environment.
3. The *ampullae of Lorenzini* open on the skin surface through pores. Further details of these structures will be found in chapter 11. Press gently around several pores and notice the exudate. The ampullae of Lorenzini are a continuation of the lateral line system (see p. 74 and chapter 11, fig. 57) and are arranged in a definite canal system around the eye and jaws (see fig. 7).
4. The mouth of the shark is on the ventral side of the head and is crescentic in shape.
5. The *spiracle* (fig. 7) is a reduced, nonrespiratory gill slit on the upper-side of the head, caudal to the eye. The spiracle may admit water to the mouth and gill chambers even though the shark's mouth is tightly closed.
6. A *spiracular valve* permits closing and opening of the external spiracular pore.
7. The *external gill slits* lie between the mouth and pectoral fin. Each slit is separated from the next by an interbranchial septum. Water entering the mouth and the spiracle passes into the pharynx, then through the internal gill slits into the gill chambers that contain the *gill lamellae*, and exits via the external gill slits (fig. 6).
8. The *endolymphatic ducts* open via two small endolymphatic pores near the midline caudal and dorsal to the eyes.
9. The *eyes* are large in sharks and partially shielded by the *upper* and *lower eyelids*. Just inside the lower lid, and continuous with it, is a strong, white membrane, the *conjunctiva*. This tough membrane covers the eyeball.
10. The corners (*canthi*) of the eyes are the rostral *nasal canthus*, toward the snout, and the *temporal canthus*, toward the spiracle.

Cloacal Region

The *cloacal aperture* lies between the pelvic fins (fig. 8). This entire region should be examined in both sexes and the differences noted. The details of the cloaca will be described later (chapter 8). The tip of the *urinary* or *urogenital papilla* may be seen just inside the cloacal aperture and abdominal pores are located caudal and lateral to the cloaca in mature sharks. A pair of fingerlike *claspers* extend caudal from the mesial edge of each male pelvic fin. In mature males a lateral spine and a ventral hook may be present near the end of the clasper. The claspers are used for the transfer of sperm to the female during mating, or copulation. At this time the contraction of a muscular siphon sac may assist the process of ejaculation of the sperm.

THE SKIN

The shark skin has small placoid scales with central spines that project through the surface of the skin. Rub your finger very lightly over the surface of the shark to detect the placoid scale spines by touch. Now examine the scales with a hand lens or dissecting microscope and compare this view with that in figure 9. If microscopes and prepared slides are available, examine sections of the dogfish skin at 100 magnification. This will allow you to identify the details of the placoid scale and of the parts of both skin layers (epidermis and dermis).

The skin consists of an outer thickness of compact cells called the *epidermis* and a deeper portion, the *dermis*. The dermis is composed of a relatively compact inner portion and an outer loose connective tissue layer with scattered *mesenchymal* (embryonic connective tissue) cells. The deeper and more compacted layer of the dermis consists principally of collagen fibers arranged in planes parallel to the skin surface and forming alternating left- and right-handed helical (spiral) patterns around the sharks body. This pattern resembles a sheath of fabric, cut on the bias, and forms a strong but flexible outer tendon that helps to transmit forces from the trunk muscles (see chapter 3) to the head and tail and thus assists in locomotor movements.

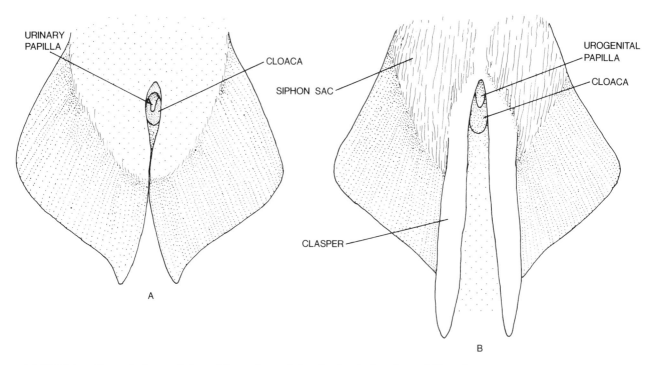

FIGURE 8. Ventral views of the pelvic regions of the shark; *A*, female and *B*, male.

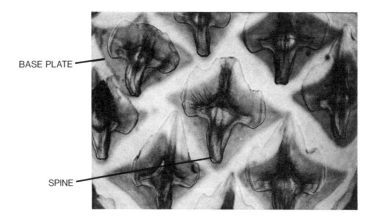

FIGURE 9. Photomicrograph of the placoid scales of the dogfish shark × 500. The skin and scale pigments have been "cleared" so both the crown and base plate are visible. The photograph is focused on the crown but the diamond-shaped base plate is visible beneath the crown. The spine projects posteriorly.

The epidermis has a basal layer (next to the dermis) of columnar cells, the *germinative layer*. From the germinative layer to the surface, the cells become progressively flatter except for a few enlarged *mucus* cells. Thus, all of the epidermal cells are living and there is no outer keratinized layer as in our own skin. This allows an osmotic exchange between the shark and its environment.

THE PLACOID SCALE

The fully developed placoid scale has a tall, caudally directed central spine and two small lateral spines at right angles to the central spine (fig. 10). This triradiate arrangement forms the *crown* of the scale. The *neck* of the scale is a pedestal supporting the crown on the *base*. The base or root of the scale is a large diamond-shaped plate with the points of the diamond corresponding to the spines of the crown. The base is held to the dermis of the skin

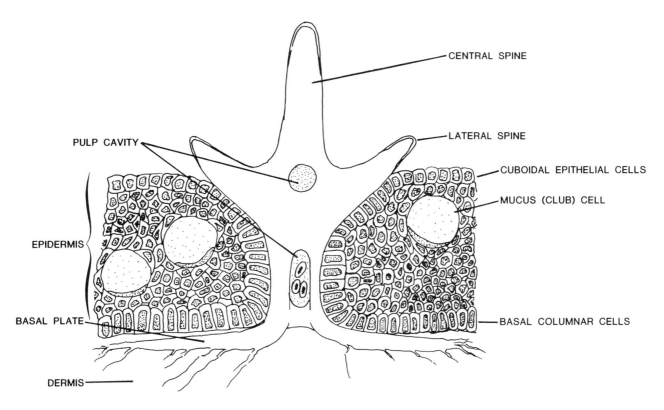

FIGURE 10. Section of the skin with a placoid scale. Note the crown, which projects above the epidermis, has two lateral and one central spine. The neck of the scale has a pulp cavity continuous with that of the base (see fig. 11). The base is anchored in the dermis by fibrous tissue.

(there is also a small amount of bone present in the base of the scale) by a thin layer of compact connective tissue and by smooth muscle fibers. Most of the dermis beneath the scale consists of dense (or compact) connective tissue. Lateral to the base of the scale is a layer of loose, areolar connective tissue containing blood vessels (or blood sinuses) and mesenchymal cells. The pulp cavity of the scale is similar to the areolar connective tissue layer.

The Development of the Scale

The first stage in scale development is indicated by an invagination of the germinative layer and an accumulation of mesenchymal cells into a papilla just beneath the invaginated epidermis (fig. 11). The epidermal cells become elongated (columnar) and arrange themselves in a cup (fig. 11). As the epidermal cup becomes better formed, the mesenchymal cells are more uniformly arranged. The mesenchymal cells constitute the papilla of the scale and the columnar germinative layer is termed the *columnar* (or *inner*) *epithelium* of the placoid scale. The edges of the columnar epithelium grow downward to enclose the papilla and become "folded" against the cuboidal layer of the germinative layer. The *cuboidal epidermal layer* adjacent to the columnar (inner) epithelium is known as the *outer epithelium*. The mesenchymal papilla is now situated directly on the more compacted collagen portion of the dermis (fig. 11).

The inner epithelium serves as a mold for the formation of the scale. There is some evidence that the inner epithelium secretes the outer hard coating (enamel) of the scale. The most popular concept is that the hard coating of the scale is the first secretion of the cells of the papilla. The papillary cells secrete the remainder of the hard tissue of the scale. The outer covering of the scale has been variously termed, enamel, mesodermal enamel, vitrodentine, and durodentine (see Teeth, p. 32). The inner hard portion of the scale is usually termed *dentine*. Before erupting through the skin, the central spine is parallel to the skin surface. During eruption the scale rotates so the spine is eventually at a 45 degree angle to the skin surface. The factors causing this rotation are unknown but it is almost identical to the rotation during tooth eruption (fig. 30, p. 33).

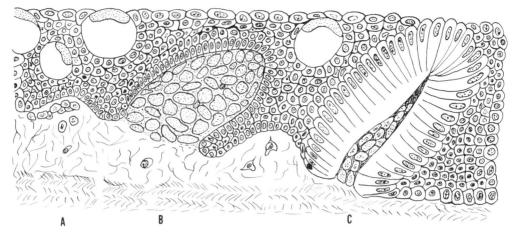

FIGURE 11. Section through the skin of the shark showing successive stages in the development of a scale. The inner (basal) layer of the epidermis serves as a mold for the scale formation. A papillae (*A*) forms in the dermis and becomes concentrated in an epidermal cup (*B*). The dentin of the scale is produced by the papillary cells and shaped by the columnar epidermal cells (*C*).

SUGGESTED READINGS

Del Rosario, Pani M., and L. O. Cimarosti. 1979. Comparative histologic study of the scales and teeth of the shark *Squalus-sp. Acta Zool. Lilloana* 35 (1): 269–72.

Nashimoto, K. 1981. The swimming speed of fish in relation to frequency of tail beating and body type. *Bull. japon. Soc. Sci. Fish.* 46 (6): 675–80.

Reif, W. E. 1980. Development of dentition and dermal skeleton in embryonic *Scyliorhinus caniculla. J. Morph.* 166 (3): 275–88.

Thomson, K. S., and D. E. Simanek. 1977. Body form and locomotion in sharks. *Amer. Zool.* 17 (2): 343–54.

Wainwright, S. A., F. Vosburgh, and J. H. Hebrank. 1978. Shark skin: function in locomotion. *Science* 202: 747–49.

Webb, P. W., and G. R. Smith. 1980. Function of the caudal fin in early fishes. *Copeia* 1980 (3): 559–62.

Webb, P. W. 1984. Body form, locomotion and foraging in aquatic vertebrates. *Amer. Zool.* 24: 107–20.

Chapter 2
Skeletal System

Shark skeletons are mainly cartilaginous, but mature specimens of certain species may have varying amounts of calcium carbonate or calcium phosphate deposition as a result of precipitation. This process is not truly bone formation because cell action that prepares the calcium hydroxyapatite molecule in the proper configuration for binding with collagen molecules does not occur. In sharks the calcium salts precipitate in the extracellular cartilage and concentrate in the gel-like cartilage matrix. The heaviest deposits of calcium are at the periphery of the cartilage, and central portions have less dense concentrations of these salts.

A checkered pattern of calcified cartilage termed *tesserae* occurs in the jaws, gill arches, vertebral arches, and fin cartilages of many sharks. These calcifications occur at the borders between the cartilage and the perichondrium, which provides chondrocytes for cartilage production. In truth, chondrocytes and osteocytes (which produce bone) are identical cells supplied with different substrates or physical factors inducing them to form either bone, cartilage, or even fibrous tissue in response to a given set of conditions.

Use considerable care when handling the prepared skeletons because they are delicate and easily broken. If you remove the skeletons from their liquid preservative do not allow them to dry. Drying shrinks and distorts the skeleton so the parts are unrecognizable.

AXIAL SKELETON

The head skeleton of the shark consists of a *chondrocranium* enclosing the brain and special sense organs (ear, eye, and nose) and a *splanchnocranium* surrounding the mouth and pharynx.

Chondrocranium

The chondrocranium appears as a single cartilaginous structure without divisions or sutures between units of the skull as seen in other vertebrates. Even the bony fishes have sutures between the bones of the skull. Of course the homologous cranium of bony fishes is constructed of bone rather than cartilage, but in addition, bony fish and other vertebrates have a second bony shell, the *dermatocranium* surrounding the *endocranium* (the bony

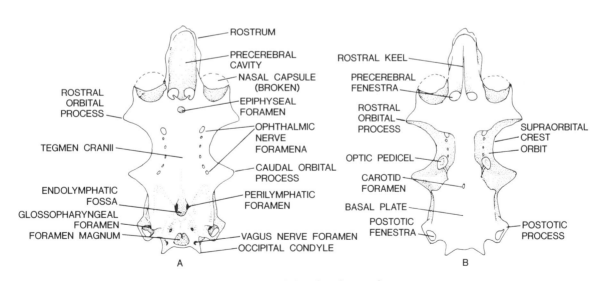

FIGURE 12. Dorsal (*A*) and ventral (*B*) views of the chondrocranium.

fish counterpart of the chondrocranium). In sharks the outer shell does not form and the chondrocranium only, is present. Nevertheless areas of the floor and walls of the chondrocranium corresponding to bones of the vertebrate skull may be identified as well as some unique parts that are not present in other vertebrates. Locate the parts illustrated in figures 12 through 14 and listed below:

1. The *rostrum* is a median cranial projection supporting the snout. Its dorsal surface is hollowed out to form the *precerebral cavity* or *rostral fossa*. Its caudal margin is bounded by two holes, the *precerebral (=rostral) fenestrae*. On its ventral surface (fig. 12B) is the *rostral carina*, a narrow keel extending from between the fenestrae forward for about half the length of the rostrum.
2. *Nasal capsules* lie on either side of the caudal end of the rostrum. These are nearly spherical as viewed from the dorsal or the ventral aspect, but they are partially open or incomplete ventrally where the external nares communicate with delicate olfactory lamellae (organ of smell, see p. 74) within the capsules. The cartilage of the capsule is very thin and is often broken on prepared crania.
3. The *rostral orbital process* projects as a nearly horizontal shelf of cartilage just caudal to the nasal capsule and beneath the rostral portion of the *supraorbital crest*.
4. A shelflike *supraorbital crest* forms the dorsal wall of the orbit (eye-socket).
5. The *epiphyseal foramen* is a small, midline opening on the dorsal side of the skull between the rostral ends of the supraorbital crests. The *epiphysis (pineal body)* projects through this foramen from the brain.
6. Several *ophthalmic nerve foramina* occur on each side, in a row, at the base of the supraorbital crest. These foramina permit passage of the superficial ophthalmic branches of the facial nerve. Additional foramina for facial nerve branches penetrate the dorsum of the nasal capsules.
7. The *tegmen cranii* is the roof of the chondrocranium.
8. A caudal *orbital process* projects laterally from the caudal end of each supraorbital crest.
9. The *endolymphatic fossa* is a midline depression caudomedial to each caudal orbital process. It is elliptical in outline and contains in its floor two pairs of foramina. The rostral pair are the *endolymphatic* and the caudal pair are *perilymphatic* foramina. These foramina open to the *endolymphatic* and *perilymphatic* ducts that connect the tissues of the fossa with the internal ear. Fluids of the ear are filtrates of seawater that are processed through the tissues of the endolymphatic fossa.
10. The *otic (auditory) capsule* is a bulky process on each side of the endolymphatic fossa. This capsule houses the internal ear and is the third of three pairs of sensory capsules associated with the chondrocranium. The others are the nasal capsules (see 2 above) and the *optic capsules* (eyes), which are protected by the orbits and adjacent processes.
11. The *foramen magnum* is a large hole through the caudal end of the chondrocranium providing an exit for the spinal cord from the chondrocranium.
12. The *occipital condyles* are a pair of articulating processes projecting caudally on either side of the foramen magnum.
13. Foramina for the *vagus nerve* are a pair of small holes lateral to each occipital condyle for the passage of the vagus nerves.
14. The *glossopharyngeal nerves* exit the cranium through *glossopharyngeal foramina* at the caudodorsal corner of the cranium.
15. The optic nerves from the eye reach the brain through *optic foramina* on the midventral border of the orbit (fig. 13).
16. The eyeball is projected outward from the orbit by an *optic pedicel,* a stalked, concave mushroom-shaped cartilage just caudal to the optic foramen (fig. 13).

Splanchnocranium

This is the visceral portion of the head skeleton supporting the mouth and pharynx (p. 32). The splanchnocranium consists of seven paired visceral arches each formed of elongated cartilaginous supports. The first two pairs of arches are associated with the jaws and the last five support the gills. Note the relationship of the

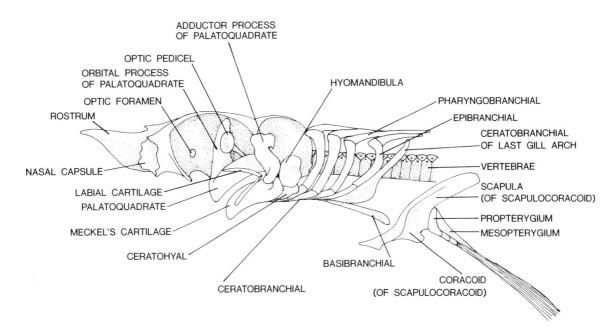

FIGURE 13. Lateral view of the cranium, jaws, branchial basket, and pectoral girdle of the shark.

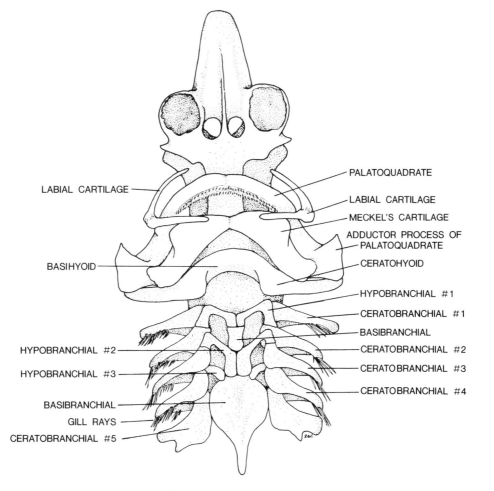

FIGURE 14. Ventral view of the cranium, jaws, and branchial basket of the shark.

splanchnocranium to the chondrocranium and to the vertebral column. Learn the location and relationships of the parts described below and illustrated in figures 13 and 14.

1. Visceral arch I, the first arch, is termed the *mandibular arch*. It includes both upper and lower jaws.
    a. The upper jaw is formed by the right and left *palatoquadrate* (=pterygioquadrate) cartilages, which are joined at their medial ends in a symphysis. These cartilages are closely attached to the ventral sides of the chondrocranium by ligaments. A prominent *orbital process* of the palatoquadrate extends into the orbit to provide a secure alignment of these cartilages with the chondrocranium.
    b. The lower jaw is formed by a pair of *Meckel's cartilages*. At their caudal ends they hinge with the palatoquadrates to form the *quadratoarticular joint* that permits opening and closing the mouth. The *hyomandibula* (part of the second visceral arch) acts as a suspensor of the lower jaw by its articulation with the otic portion of the chondrocranium above, and with Meckel's cartilage below.
    c. In addition to the palatoquadrate and Meckel's cartilages, paired slender *labial cartilages* help support the lips and aid in strengthening the angles and sides of the jaws.
    d. Teeth are prominent structures derived from the dermis of the skin covering the biting portions of both upper and lower jaws. Notice their arrangement in rows, one behind the other. Lost teeth from the outer rows are replaced as teeth from behind move forward (see p. 32, chapter 5, and fig. 30).
2. Visceral arch II, the *hyoid arch*, lies immediately caudal to the first arch but is less robust than the first.
    a. Ventrally this arch consists of a median *basihyal cartilage* from which paired *ceratohyals* extend dorsolaterally as slender bars. These three cartilages support the floor of the mouth.
    b. Bridging the space between the dorsal end of each ceratohyal and the otic capsule, on either side, is a small *hyomandibular cartilage,* also bar-shaped, called the *hyomandibula*. Thus five cartilages, the unpaired basihyals, together with the paired ceratohyals and paired hyomandibulars form the second arch.
    c. Jawless vertebrates (Agnatha) have nine or ten pairs of functional gill arches. In the most primitive jawed vertebrates (Gnathostomes; see fig. 1) the first pair of gill (pharyngeal) arches of the embryo develops into a simple, small but efficient, pair of jaws. Six or seven pairs of functional gill arches remain as exemplified in the primitive living sharks (*Hexanchus* and *Heptanchus*) that have six and seven pairs of gill slits, respectively. *Squalus acanthias,* a more specialized shark, has large jaws formed of the first two pairs of pharyngeal arches and five pairs of gill chambers supported by functional gill arches (figs. 13 and 14). This progressive series nicely represents a range of variation from a jawless suctorial feeding mouth to the strong, wide biting mouth of most sharks and bony fishes (teleosts). The species with strong, wide mouths are usually the most active and voracious feeders.
3. Visceral arches III–VII are the *branchial arches*. These are dorsally incomplete "hoops" bearing gill lamellae only in their midlateral regions and with dorsal and ventral cartilages serving for attachment. Ventrally, the right and left branchial arches are attached through two of three unpaired *basibranchials* and three pairs of *hypobranchials* (fig. 14). The caudal basibranchials may be fused or separate. Together these cartilages resemble a paddle with the "blade" portion between the last two pairs of *ceratobranchials* and the "handle" portion directed caudally.

From rostral to caudal on the ventral pharynx, the first pair of hypobranchials attach to the first basibranchial, which attaches to the second pair of hypobranchials. The third pair of hypobranchials are lateral to the second pair, and both the second and third pairs of hypobranchials attach caudally to the second basibranchial. The first three pairs of ceratobranchials are jointed to the three pairs of hypobranchials. The last two pairs of ceratobranchials are jointed to the second basibranchial (fig. 14).

The ceratobranchials are also attached on their dorsal end to *epibranchials* and each epibranchial is attached dorsally to a *pharyngobranchial*. The pharyngobranchials extend caudally and medially from their attachment with the epibranchials but they are held in place only by connective tissues and there is no direct attachment to the vertebral column.

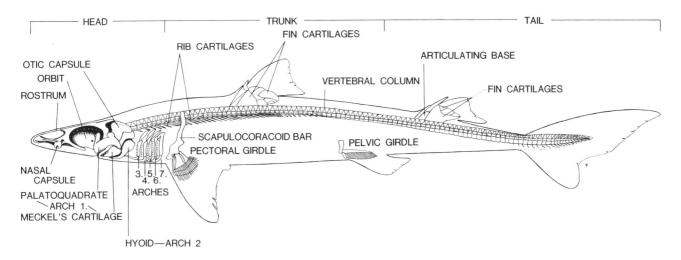

FIGURE 15. Diagrammatic lateral view of the entire shark skeleton.

VERTEBRAL COLUMN

The vertebral column is composed of two major types of vertebrae: the body (or trunk) vertebrae and the nearly symmetrical tail vertebrae.

1. *Tail vertebrae*—Study the details of a tail vertebra on a prepared skeleton and, again on your dissection specimen, where the tail has been freshly cut. Identify all of the vertebral structures shown in figure 16. Notice the *neural canal,* which shelters the spinal cord, the *haemal canals* surrounding the caudal artery and vein, the *neural spine* projecting dorsad from the *neural arch* and the *haemal arch* projecting ventrad from the *haemal arch.* Note also the remains of the *notochord* lying in the vertebral centrum. Make several thin transverse sections across the tail close to your first cut, and observe the varying thicknesses of the notochord. The diameter of the notochord depends upon whether the section passes through the middle or near the end of the vertebra. The tail vertebrae are laterally compressed and angle slightly upwards (fig. 15). See the description of the tail on pages 2–4.
2. *Body (trunk) vertebrae*—These are similar to the tail vertebrae but they lack a haemal arch. The base of a ventral arch is present, however in the form of two small cartilages or *basapophyses* (basal stumps). Locate the cartilaginous *ribs* that extend into the *transverse septum* from the basapophyses (=parapophyses) with which they articulate. Identify also the *intercalary plates* (arches) interspersed like wedges between the neural arches of the vertebrae (fig. 16A,B). The *corpus calcareum vertebrae* is the most calcified part of the vertebral centrum.

APPENDICULAR SKELETON

Pectoral Girdle

Identify the various parts illustrated in figures 13 and 17 on a preserved pectoral girdle. The *scapular process* and the slender *suprascapular cartilage* curve dorsally from the junction of the scapula with the thick *coracoid* portion that arches ventromedially to unite with its fellow of the opposite side. The united coracoids form the *coracoid bar* (coracoid arch) of the pectoral girdle. The *glenoid surface* at the junction of the scapular process and coracoid provides for the articulation of the pectoral fin with the pectoral girdle. Strong *basal fin cartilages,* the *pro-, meso-,* and *metapterygia* articulate at the glenoid surface. Many slender *radial cartilages* extend from the basal cartilages into the fin to provide rigidity while the dermal *ceratotrichia* allow flexibility in the distal part of the fin. Pro-, meso-, and metapterygia together with the radials are collectively termed *pterygiophores* (Greek = wing or fin bearers).

Pelvic Girdle

Identify the structures of a pelvic girdle noting the difference between the sexes in mature sharks. In the male, each pelvic fin bears a clasper on its medial side (fig. 17). The clasper is supported by the largest radial cartilage. The fin illustrated is from a small male and the clasper is somewhat immature. Note the absence of a mesopterygium in the pelvic girdle. Compare claspers of several male sharks when opportunity affords.

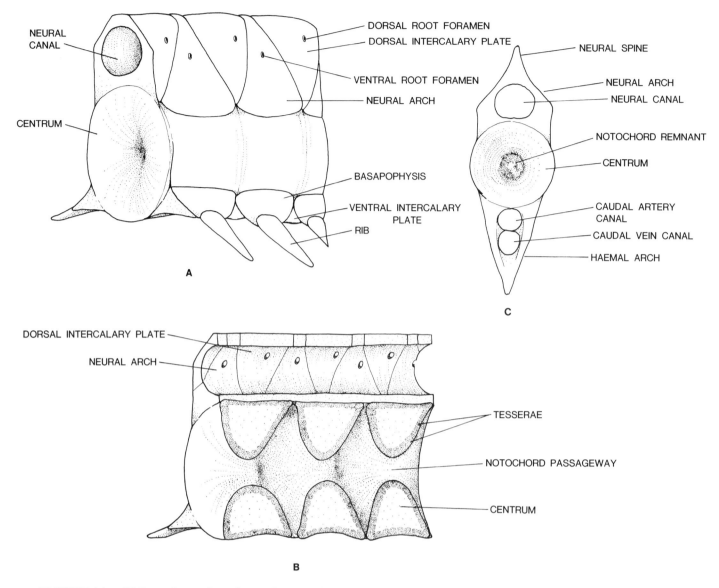

FIGURE 16. Oblique views of trunk vertebrae, entire (*A*), sectioned (*B*) and frontal views (*C*) of a caudal vertebra.

If separate skeletons of shark girdles are not available, you should strip off the skin over the two girdles on the shark's left side and from the bases of the fins, being careful to leave the muscles intact. In order to find the scapular process of the the pectoral girdle, you will have to separate the skin from the muscles dorsal to the cranial edge of the fin. After the muscles have been studied, a more detailed examination of the fin cartilages should be made.

SUGGESTED READINGS

DuBrul, E. L. 1964. *The temporomandibular joint*. 2d ed. Chapter 1. Springfield, Ill.: Charles C Thomas Publisher.

Edgeworth, F. H. 1926. On the hyomandibula of Selachii, Teleostomi, and Ceratodus. *J. Anat.* 60.

Kemp, N. E., and S. K. Westrin. 1979. Ultrastructure of calcified cartilage in the endoskeletal tesserae of sharks. *J. Morph.* 160 (1): 75–102.

Maisey, J. G. 1980. An evaluation of jaw suspension in sharks. *Amer. Mus. Novitates* no. 2706, pp. 1–17.

Peignoux-Deville, J., F. Lallier, and B. Vidal. 1981. Evidence of the presence of osseous tissue in dogfish vertebra. *C.R.Seances Acad. Sci. ser.* III, sci. vie 292 (1): 73–78.

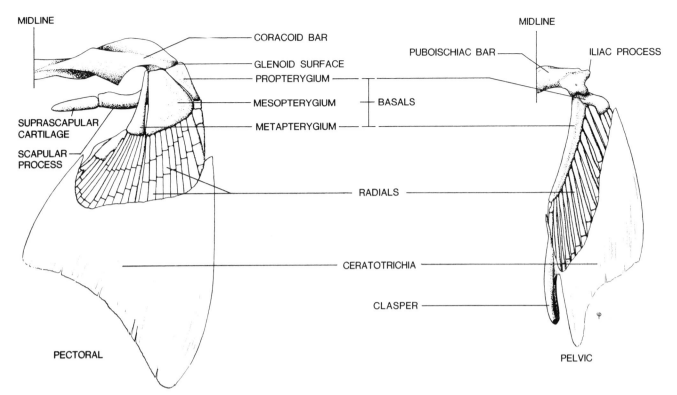

FIGURE 17. Ventral views of the left halves of the pectoral (left) and pelvic (right) girdles and limbs of the male shark.

Chapter 3
Muscular System

Vertebrate muscle fibers are of three types: (1) striated, (2) smooth, and (3) cardiac. Striated muscles are used in swimming, feeding, and other behavioral activities; smooth muscles line the digestive tract and blood vessels; and cardiac muscle fibers are restricted to the heart.

Striated fibers are further arranged as "red" or "white" with several intermediate types. These fibers have not been well studied in fish but the dogfish, *Scyliorhinus canicula,* is known to have at least four fiber types. The first, the outer red (OR) fibers have a greater succinic dehydrogenase but a lesser ATPase content than the inner red (IR) fibers. Red fibers contract more slowly and are relatively nonfatigable compared to white fibers. The red fibers are involved in moderate but constant activity such as posture maintenance in tetrapods or slow swimming in fishes.

White fibers are also grouped as outer (OW) and inner (IW) fibers. White fibers are faster contracting but require less oxygen than red fibers. These muscles are used for fast movement such as rapid swimming.

Red fibers require a rich blood network to deliver a constant oxygen supply necessary for oxidation of the lactic acid produced during contraction. In the dogfish the OR fibers require the richest supply of blood vessels. This intense vascular bed may be arranged as a "rete mirable" with the venous blood flowing in one direction and arterial blood flowing in the other. This countercurrent mechanism in some sharks and large fishes serves to exchange heat from the veins warmed by the contracting muscles to the cooler arteries that carry blood from the gills. In this way the shark is able to maintain a body temperature four or five degrees Celsius above the environmental water temperature.

BODY MUSCULATURE

Before beginning your dissection, cut each of the dorsal spines of your specimen so you do not snag your gloves or puncture your hand while doing the rest of your dissection.

Follow the broken lines in figure 18 and cut (1) through the body wall a little to the left (of the shark) of the midventral line starting at the left side of the cloaca and continue forward to the mid lower lip. Next, cut (2) through the skin from midventral to middorsal lines at a point just cranial to the pelvic fins and again (3) just caudal to the pectoral fin as in figure 18. Grasp the skin on the left side at the midventral incision, and peel or tear it dorsad to expose the muscles. Some dissection may be required in removing the skin from this area. Cut (4) just cranial to the pectoral fin and remove the skin on the ventral surface of the pectoral fin. Next, cut (5) the skin just caudal to the lower lip, dorsally just caudal to the spiracle to the top of the head. Grasp the edges of the skin and peel it away from the muscles. Be especially careful removing the skin over the gill arches. The thin and delicate muscles of the gill arches are easily torn loose with the skin. Scrape these muscles from the inner surface of the skin as you pull the skin loose in this area. Be sure to rewrap your shark with a wet paper towel before putting the specimen away in the plastic bag.

Observe and identify the muscles and septa exposed by these dissections.

1. *Myotomes* are the individual muscle segments of the shark trunk. Each myotome appears as a zigzag segment from the dorsal midline to the ventral midline. The muscle fibers making up each myotome are directed horizontally (i.e., from rostral to caudal).
2. *Myosepta* (=myocommata) are connective tissue sheets separating myotomes from each other. The myotome muscle fibers insert on myosepta.
3. *Transverse septum* is a horizontal sheet of connective tissue dividing each myotome into dorsal (epaxial) and ventral (hypaxial) portions. The *lateral line* organ lies just lateral to the edge of the transverse septum thus providing a superficial "landmark" for the septum.
4. *Epaxial* muscles, above the transverse septum are collectively homologous to the dorsalis trunci muscle mass of the tailed amphibians (fig. 20).
5. *Hypaxial* muscles, ventral to the transverse septum are homologous to the trunk muscles of tetrapod vertebrates. These muscles may be more or less separable into three groups or bundles. Muscle fibers just ventral to the transverse septum

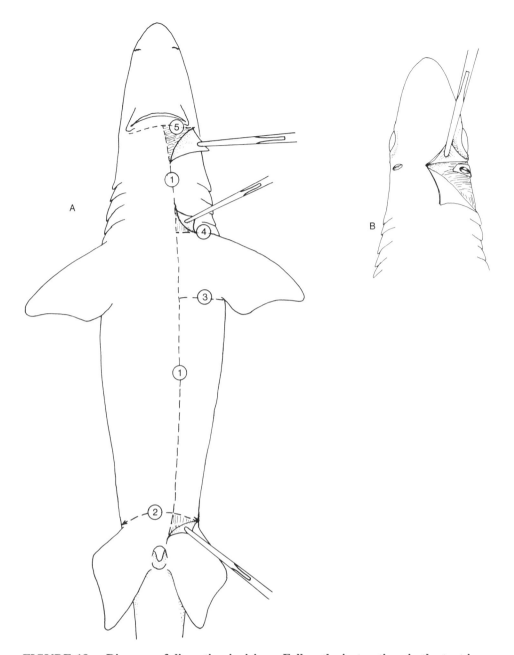

FIGURE 18. Diagram of dissection incisions. Follow the instructions in the text in making incisions. Cut through the *skin only* and use a blunt probe to separate the skin from the underlying muscles. *A*, is a ventral view of the shark, and *B*, is a dorsal view of the head.

and extending ventrally through the next complete bend of the myotome (fig. 20) make up the *lateral bundle*. The lateral bundle muscles are homologous to the subvertebral, internal abdominal oblique, and transversus abdominis muscles of tetrapod vertebrates. The remaining ventral portions of the myotome are the *ventral bundle* and *rectus abdominis*. The rectus abdominis muscles are the portions of the myotomes nearest the ventral midline of the body. These muscles are thinner than the ventral bundle muscles and are separated from each other by a tough connective tissue sheet in the ventral midline called the *linea alba*. The ventral bundle is homologous to the external abdominal oblique muscles of tetrapod vertebrates.

6. *Medial septa* separate the right and left muscle masses above and below the vertebrae in the tail and above the vertebrae in the trunk region. The septa above the tail vertebrae is the *dorsal median septum* and that below the tail vertebrae is the *ventral median septum*. Identify these structures on a cross section of the shark tail (fig. 19).

17

Myotome Contraction

Each bend of a myotome is related to the corresponding point of the preceding or succeeding myotome like one of a stack of paper cups is related to a preceding or succeeding cup (see fig. 20). That is, the myotome does not extend directly from the midline to the surface, but angles, cranially on the dorsal and ventral points, and caudally on the middle point. Thus, if a horizontal section is cut so the myotome is divided at any of these bends, we would find the muscle segment shaped as in figure 21.

Each muscle fiber is arranged parallel to the midline and extends from one myoseptum to the next (see fig. 21). The simultaneous contraction of all the fibers at this level tends to move the two myosepta toward one another. This actually may happen to a slight extent but several factors work against it.

If the myoseptum were to move to a position parallel to its resting position, then any point on the myoseptum may have a movement vector perpendicular to the surface of the myosepta. We may indicate this as a vector of the force of the contracting muscle fiber that acts with a second vector, 90 degrees from the first, to move the myosepta. The vectors on either end of the myotome that are perpendicular to their respective myosepta surfaces will cancel each other out because they are equal in strength and directly opposed. The effective vectors are parallel to the myosepta surfaces and tend to torque the myotome (see fig. 21).

If a caudal-medial myotome segment is opposed to a cranial-dorsal (or ventral) myotome segment, the opposing torques will bend the midline of the body at a point midway between the two myotomes.

A second consideration of myotome contraction concerns shape changes. Each muscle cell or fiber contracts one-half its length. When a single cross-sectional area of muscle fibers contract, the cranial-caudal distance is reduced by one-half. Since the volume does not change, the myotome must expand laterally. The slanting arrangement of the myotome presents less resistance to the lateral expansion of the muscle fibers than would a perpendicular segment with parallel fibers. Since the increasing torque of the myotome increases the degree of slant, the resistance to lateral expansion is reduced by the act of contraction.

A third factor in myotome contraction involves the sequence of innervation of the myotomes. Muscle contractions are initiated by nerve impulses originating in the shark brain and traveling caudally in nerves of the spinal cord. From the spinal cord, spinal nerves branch to the lateral myotomes. There is a spinal nerve serving each myotome so each myotome contracts in order from cranial to caudal. All the fibers in each myotome contract simultaneously but the slanted arrangement of the myotomes provides for a smooth increase and decrease in contraction strength at each body level.

The complex arrangement of fish myotomes is due to three selective factors: (1) the ability to torque and thus

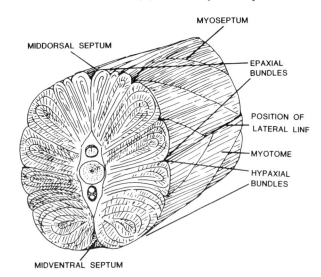

FIGURE 19. Oblique view of a cross section of the tail of the shark illustrating the relationships of muscle bundles to myotomes.

produce a bending movement, (2) the necessity of the segment to expand laterally as it contracts cranial-caudally, and (3) the innervation and overlapping of myotomes produces a graded increase and decrease in contraction at each cross-sectional level of the body.

BRANCHIAL AND HEAD MUSCLES

After you have cleaned off the skin over the gill slits, cut around the spiracle and grasp the edge of the skin with hemostats or locking forceps as in figure 18B and pull the skin of the head to expose the brachial musculature.

The dorsal bundle of the epaxial myotomes extends cranially, above the gill slits and attaches on the caudal surface of the chondrocranium. The lateral bundle of epaxial myotomes attaches to the scapular cartilage. Below the transverse septum, the lateral and ventral bundles of hypaxial muscle attach cranially to the coracoid bar.

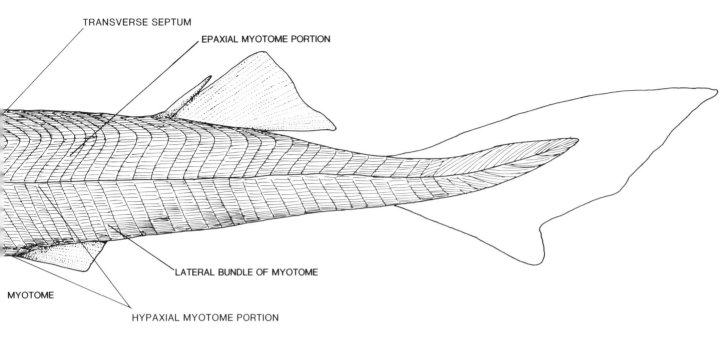

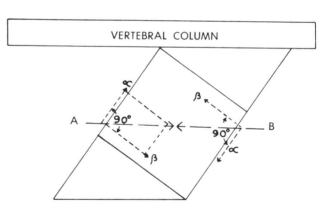

FIGURE 21. Diagram of myomere movement. Since each myoseptum is drawn toward the other, A and B indicate the forces in the myomere that draw the myosepta together. These forces are produced by the contraction of the muscle fibers. One of the vectors (β) of each of the forces A or B may be perpendicular to each of the myosepta. The vectors are equal but opposed to similar vectors of other myofibrils and are canceled out. The other vectors (α) for forces A and B are parallel (90° to the first vectors) to the surface of the myosepta. Since they are not directly opposed, these are the effective forces in the movement of the myomere. Actually the set of vectors tends to torque the myomere in a direction opposite to the set. The β set of vectors are ineffective except at the two extremes of the myomere. The resultant movement at any given level is a torque of the myomere caused by the parallel α vectors. (From *Laboratory Anatomy of the Perch,* Wm. C. Brown Company Publishers.)

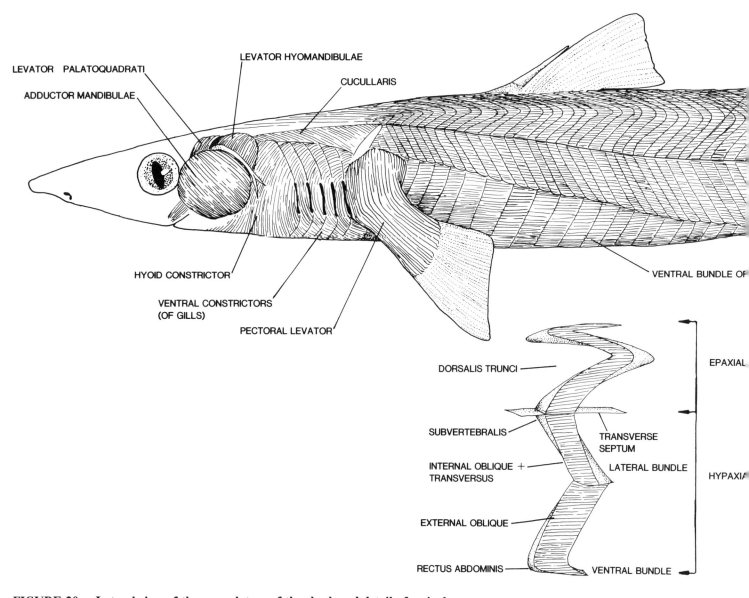

FIGURE 20. Lateral view of the musculature of the shark and detail of a single myotome. Labels to the left of the myotome are homologous amphibian (tetrapod) muscles.

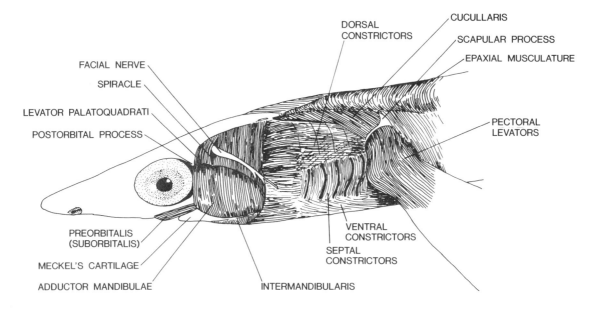

FIGURE 22. Lateral view of the head and branchial muscles.

The *branchial muscles* reach from the coracoid bar to the mouth and to the eye. They provide the movements needed to open and close the mouth and the gill arches and are therefore involved in feeding and respiration. These muscles occur in series, corresponding to the serially arranged branchial arches. They include *constrictors* that act to close the mouth, compress the gill chambers, and assist in swallowing; *levators* or "lifting" muscles that raise the jaws, lift the hyoid arch, and elevate the gill arches and scapular process; and *hypobranchial* muscles that help to open the mouth and assist in swallowing.

1. *Constrictor series*—Six *dorsal* and six *ventral superficial constrictors,* above and below the gills, make up this series. First are the constrictors of the mandibular arch; second, those of the hyoid arch; and finally, the branchial constrictors of the gill arches (branchial arches 3–6). The last four constrictor groups (sets) reveal only their caudal ends since each is mostly overlapped by the one in front (figs. 20, 22, 23, 24, and 25).

   NOTE: Branchial muscles, like most body muscles, are paired and do not reach across the midline. Each muscle has a more fixed (immovable) *origin* and a more movable *insertion.*

   a. A white vertical connective tissue band, the *raphe,* extends above and below each gill slit and separates adjacent constrictor sets. Cut along the raphe upward and downward from one gill slit, thus exposing the concealed portion of one of these sets of superficial constrictors.
   b. The *dorsal constrictor* of a gill arch originates on the fascia of dorsal myomeric muscle and inserts on superficial tissue of the gill septum. Gill constrictors compress the parabranchial cavity (fig. 22).
   c. The *ventral constrictor* of a gill arch originates on ventral myomeric fascia and inserts on superficial tissue of the gill septum.
   d. The innermost part of the constrictor forms the *septal constrictor* muscle. These consist of deep muscle fibers in the gill septum between the dorsal and ventral constrictors. The mandibular and the hyoid constrictor sets are more complex.
   e. The *adductor mandibulae* is the main muscle of the first set. Its origin is on the caudal margin of the palatoquadrate cartilage. It inserts on the caudal lateral surface of Meckel's cartilage and *adducts* (closes) the jaws.
   f. The *preorbitalis* (=suborbital) muscle below the orbit is cylindrical in shape, originates on the ventral caudal midline of the rostrum, and inserts on the border of Meckel's cartilage just caudal to the *quadrato-articular* joint. It acts to expand the oral cavity and assists in opening the jaws. Deep dissection is needed to find it.
   g. A disputed dorsal component of the first constrictor series, the *spiracular* (craniomaxillary) courses dorsal-ventrally just rostral to the spiracle and is difficult to separate from the palatoquadrate levator in front of it. Both of these muscles (if indeed they are separate) originate on the otic capsule of the cranium and insert dorsally on the caudal part of the palatoquadrate cartilage.

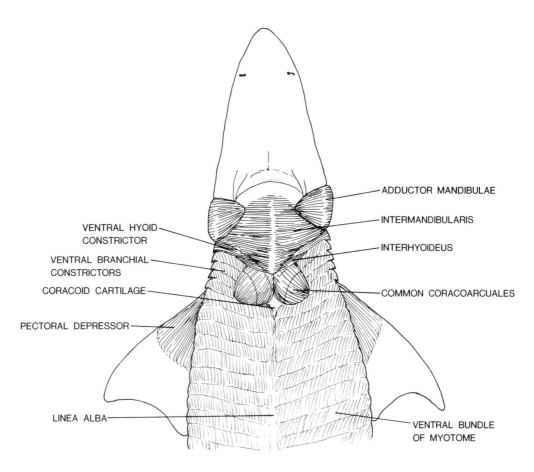

FIGURE 23. Ventral view of the superficial jaw and branchial muscles.

h. A large, thin constrictor, the *intermandibularis*, originates on the ventral margin of Meckel's cartilage and inserts at the midline on its fellow of the opposite side. It acts to compress the jaws in opposition to the preorbitalis. This muscle is difficult to separate from the interhyoideus (see j below). Before cutting, use a blunt probe to separate this muscle from the interhyoideus. Insert the probe between the two muscles near their origin on the mandible and carefully move the probe horizontally separating the two layers. Note the slight differences in fiber direction of the two layers.

i. The second, *hyoid* constrictor set has a large *dorsal hyoid constrictor* whose origin is on the dorsal myomeric muscles. Its insertion is on the superficial tissues of the gill septum or raphe. It acts to constrict the parabranchial cavity.

j. Ventrally, a thin hyoid constrictor called the *interhyoideus* is just dorsal to the intermandibularis. The interhyoideus and intermandibularis are extremely difficult to separate from each other.

2. *Levator series*—These are two large dorsal muscles found both in front of, and behind, the spiracle (figs. 22 and 25).

 a. The *levator palatoquadrati* is rostral to the spiracle and originates on the rostral lateral wall of the *neurocranium* (otic capsule) and inserts on the dorsocaudal part of the palatoquadrate cartilage. The caudal portion of this muscle may be separable as the spiracular (see 1, g above).

 b. The *levator hyomandibuli* is caudal to the spiracle and overlaps the rostral margin of the dorsal hyoid constrictor. This muscle originates on the lateral wall of the neurocranium and inserts on the lateral surface of the hyomandibular cartilage. Both levators act to raise the jaws.

 c. A possible third levator, the *cucullaris,* may represent the remaining levators of the gill arches, but its origin on the dorsal occipital surface of the neurocranium and the cranial epaxial muscle fascia and its insertion on the scapular part of the scapulocoracoid strongly (Text continues on page 24.)

CONSTRICTOR SERIES

NAME	ORIGIN	INSERTION	ACTION
Adductor mandibulae	Caudal palatoquadrate	Caudal-lateral surface of Meckel's cartilage	Adducts (closes) jaws
Preorbitalis (=suborbitalis)	Ventral-caudal midline of rostrum	Border of Meckel's cartilage caudal to quadratoarticular joint	Expands jaws and oral cavity. Assists in opening mouth
Intermandibularis	Ventral margin of Meckel's cartilage	On its fellow in the ventral midline	Compresses jaws in opposition to preorbital
Constrictor hyoideus	Fascia of dorsal myomeric muscles and otic capsule	Hyomandibula and superficial tissues of gill septum	Constricts rostral end of gill chamber
Interhyoideus	Midventral raphe	Ceratohyal cartilage	Compresses gill chambers and helps raise floor of mouth
Dorsal constrictors	Fascia of dorsal myomeric muscle	Superficial tissue of gill septum	Constricts parabranchial cavity
Ventral constrictors	Ventral myomeric muscle fascia	Superficial tissue of gill septum	Constricts parabranchial cavity
Septal constrictors	Deep muscle fibers in the gill septum between and continuous with the dorsal and ventral constrictors		

LEVATOR SERIES

NAME	ORIGIN	INSERTION	ACTION
Levator palatoquadrati (+ Spiracularis)	Rostral-lateral wall of neurocranium	Dorsal-caudal region of palatoquadrate	Raises jaws
Levator hyomandibuli	Caudal-lateral wall of neurocranium	Lateral surface of hyomandibula	Raises jaws
Cucullaris	Fascia of cranial epaxial muscles and dorsal occipital neurocranium	Scapular part of scapulocoracoid	Draws pectoral girdle and limb craniodorsally

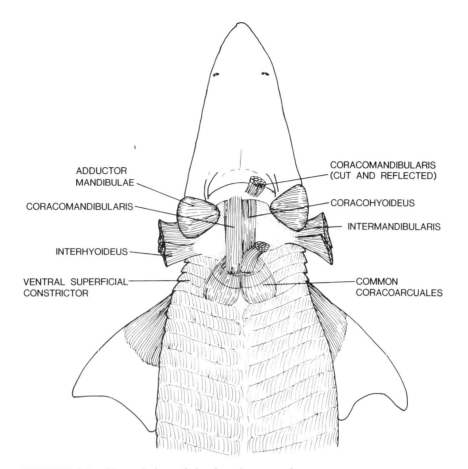

FIGURE 24. Ventral view of the deep jaw muscles.

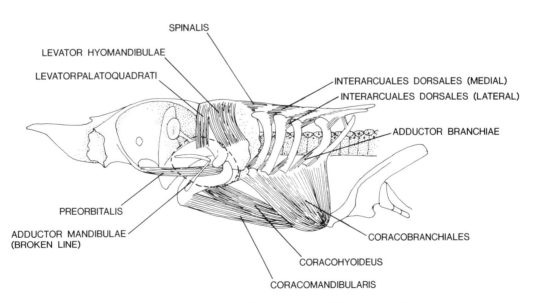

FIGURE 25. Diagrammatic lateral view of the deep branchial muscles. In part after Hughes & Ballintijn, 1965. *J. Exp. Biol.* 43:363–83.

suggests a major function of moving the pectoral girdle and fin cranially and dorsally. Therefore, it is usually considered an appendicular muscle.

3. *Interarcual series*—Expose, above the gill pouches, a large channel (the anterior cardinal sinus, see p. 51, and fig. 43) by separating the dorsal epaxial bundle and the cucullaris from the upper ends of the gill pouches. The pharyngobranchial cartilages (fig. 13) will be found in this dissection and should be spread apart to find the following:
   a. The short *dorsal interarcual* muscles originate on the ventrocaudal half of the first, second, and third pharyngobranchial cartilages and insert on the dorsocranial half of the second, third, and forth pharyngobranchial cartilages. They help the subspinal (see c. below) in drawing the gill arch skeleton forward (fig. 25).
   b. *Lateral interarcual* muscles may also be found. Their dorsal ends lie beneath the dorsal interarcuals, and their fibers run in a more vertical direction.
   c. The *spinalis* muscle originates near the foramen magnum on the exoccipital region of the cranium and inserts on the dorsal-cranial half of the first pharyngobranchial cartilage. It acts with the dorsal interarcuals to draw the gill arches forward (fig. 27).
   d. By careful dissection down to the center of a branchial arch, you may find the small *branchial adductor*. These little muscles originate on the ventral-cranial halves of the epibranchial cartilages and insert on the dorsal-cranial halves of the ceratobranchial cartilages. They act to flex the branchial arches (figs. 25 and 31B).
   NOTE: Branchial adductors are properly termed *intraarcuals* rather than interarcuals since they lie within (intra) rather than between (inter) the branchial arches.

4. *Hypobranchial muscles*—These muscles lie between the coracoid bar and Meckel's cartilage and just dorsal to the thin intermandibularis and interhyoideus muscles of the ventral constrictor series (fig. 25).
   a. The *coracoarcualis* forms a massive muscle that originates on the ventral surface of the scapulocoracoid. The three following muscles have their origin from the cranial border of the coracoarcual.
   b. The *coracomandibularis* lies just above the interhyoideus and the intermandibularis. It inserts on the caudal-ventral surface of Meckel's cartilage and acts to open the lower jaw (figs. 23 and 24).
   c. The *coracohyoideus* is just above the coracomandibularis. It inserts on the ventral surface of the basihyal cartilage and assists the coracomandibularis in opening the jaw.
   d. The *coracobranchialis* is above the coracohyoideus. It has several slips that fan out to insert on the ceratobranchial and basibranchial cartilages of the various gill arches. The rostral-most slip may insert on the ceratohyal cartilage. Coracobranchial muscles act to expand the pharyngeal cavity.

INTERARCUAL SERIES

NAME	ORIGIN	INSERTION	ACTION
Subspinalis	Exoccipital region of cranium near foramen magnum	Dorsal-cranial one-half of first pharyngobranchial cartilage	Draws dorsal gill arch skeleton forward
Interarcualis dorsalis	Ventral-caudal half of first, second and third pharyngobranchial cartilages	Cranial-dorsal half of second, third and fourth pharyngobranchial cartilages	Assist subspinalis in drawing gill arch skeleton forward
Adductor branchiae	Ventral-cranial half of all epibranchial cartilages	Dorsal-cranial half of all ceratobranchial cartilages	Flexes branchial arches

HYPOBRANCHIAL MUSCLES

NAME	ORIGIN	INSERTION	ACTION
Coracomandibularis	Coracoid via common coracoarcual	Rostral medial surface of Meckel's cartilage	Depresses (opens) lower jaw
Coracohyoideus	Coracoid via common coracoarcual	Ventral surface of basihyal	Assists in depression of lower jaw
Coracobranchiales	Coracoid via common coracoarcual	Ceratobranchial and basibranchial cartilages (hypobranchial)	Expand pharyngeal cavity

Coracoarcualis (communis) is a large muscle mass arising from the ventral surface of the scapulocoracoid and giving rise to the coracomandibularis, coracohyoideus, and five coracobranchialis muscles.

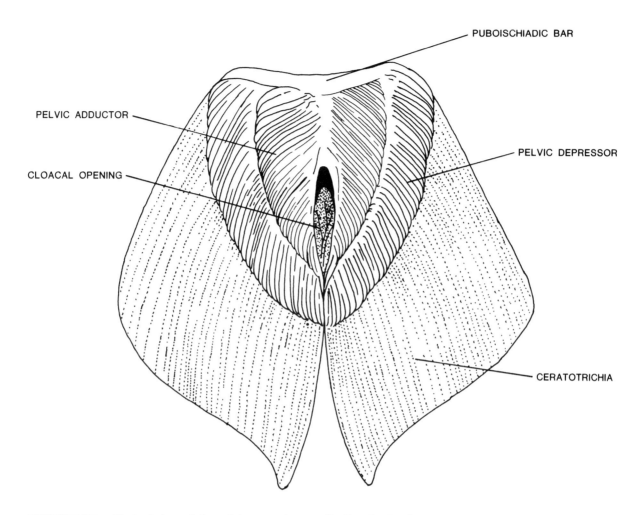

FIGURE 26. Ventral view of the pelvic musculature of a female shark.

FIN MUSCULATURE

Tear or cut the skin away from each side of a pectoral and a pelvic fin (a female is preferred since the claspers complicate the muscle pattern). Notice the fin muscles, their origins, and their insertions (figs. 20, 22, 23 and 26).

1. The *pectoral levator* on the dorsal side of the pectoral fin originates on the scapular process and adjacent fascia. It inserts on the basal and radial pterygiophores. It elevates the fin.
2. The *pectoral depressor* on the ventral side of the fin originates on the coracoid bar. It inserts on the pterygiophores of the fin and acts to depress the pectoral fin.
3. The *pelvic levator* on the fin's dorsal side takes origin on myotomal fascia and the iliac process. It inserts on the fin's pterygiophores and ceratotrichia and acts to elevate the fin.
4. The *pelvic adductor* lies next to the pelvic levator but is closer to the midline. Its origin is on the linea alba and puboischiadic bar. Insertion is on the metapterygium. This muscle adducts the fin drawing it toward the midline.
5. The *pelvic depressor* originates on the ventral side of the metapterygium and inserts on the pterygiophores and ceratotrichia of the fin. It acts to depress the pelvic fin.

APPENDICULAR MUSCLES

NAME	ORIGIN	INSERTION	ACTION
Pectoral levator	Scapular process and adjacent fascia	Basal and radial fin cartilages	Elevates pectoral fin
Pectoral depressor	Coracoid bar	Pterygiophores of fin	Depresses pectoral fin
Pelvic levator	Myotomal fascia and iliac process	Pterygiophores and ceratotrichia of fin	Elevates pelvic fin
Pelvic adductor	Linea alba and puboischiadic bar	Metapterygium	Adducts pelvic fin
Pelvic depressor	Metapterygium	Pterygiophores and ceratotrichia of fin	Depresses pelvic fin

SUGGESTED READINGS

Hughes, G. M., and C. M. Ballintijn. 1955. The muscular basis of the respiratory pumps in the dogfish (*Scyliorhinus canicula*). *J. Exp. Biol.* 43: 363–83.

Johnston, I. A., 1983. Dynamic properties of fish muscle. Chapter 2 in *Fish biomechanics*. ed. W. Webb and D. Weihs. New York: Praeger Publishers.

Satchell, G. H., and D. J. Maddalena. 1972. The cough or expulsion reflex in the Port Jackson shark, *Heterodontus portus-jacksoni*. *Comp. Biochem. Physiol.* 41A: 49–62.

Szarski, H. 1964. The function of myomere folding in aquatic vertebrates. *Bull. de L'Acad. Polonaise des Scienses.* cl. II (7) 12: 305–6.

Willemse, J. J. 1959. The way in which flexures of the body are caused by musculature contractions. *Koninkl. Nederl. Akad. Wetensch. Proc.* 62:589–93.

Willemse, J. J. 1966. Functional anatomy of the myosepta in fishes. *Koninkl. Nederl. Akad. van Wetensch. Amsterdam Proc.* series C. (1): 69:58–63.

Chapter 4
Membranes, Mesenteries, and Coelomic Cavities

DISSECTION INSTRUCTIONS

Using figure 27 as a guide, cut through the body wall to the left of the ventral body midline (1) from the cloaca to just caudal to the lower jaw. This will include cutting through the pelvic and pectoral girdles, which may be accomplished with scissors. Be especially careful when cutting the girdle cartilages that you do not cut too deep and damage visceral structures. Make a second cut transversely (2) just anterior to the pelvic girdle and a third cut transversely across the midbody region (3).

To expose the heart, cut transversely, caudal to the coracoid bar from your first cut to the right side (4), then cut forward (5) to meet cut (1) at the caudal border of the lower jaw. Remove the wedge-shaped piece of the ventral wall of the pericardial chamber to expose the heart. Your dissection should be similar to figure 32. The flaps formed by your cuts may be pinned to the dissection board with wooden handled dissecting needles.

GENERAL VISCERA

Most of the visceral organs are suspended in the body cavity by thin membranes. You should locate these major organs and their membranes before examining them as part of their respective organ systems.

The *liver* is the large grayish green organ filling most of the body cavity. The *falciform ligament* is an arched membrane connecting the cranial end of the liver with the ventral body wall and divides the right and left lobes of the liver. The large J-shaped stomach lies just medial to the left lobe of the liver and (in the female) the *uterus* lies in a similar position medial to the right lobe. The paired gonads (*testes* in the male and *ovaries* in the female) lie on either side of the dorsal midline deep to the stomach and uterus (see figs. 44 and 46). The *spiral intestine* continues posteriorly from the stomach to the *cloaca*.

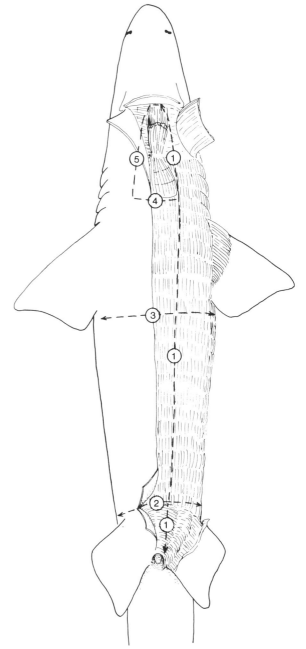

FIGURE 27. Diagram of dissection incisions. Follow the instructions in the text and be careful to *not cut* the underlying viscera.

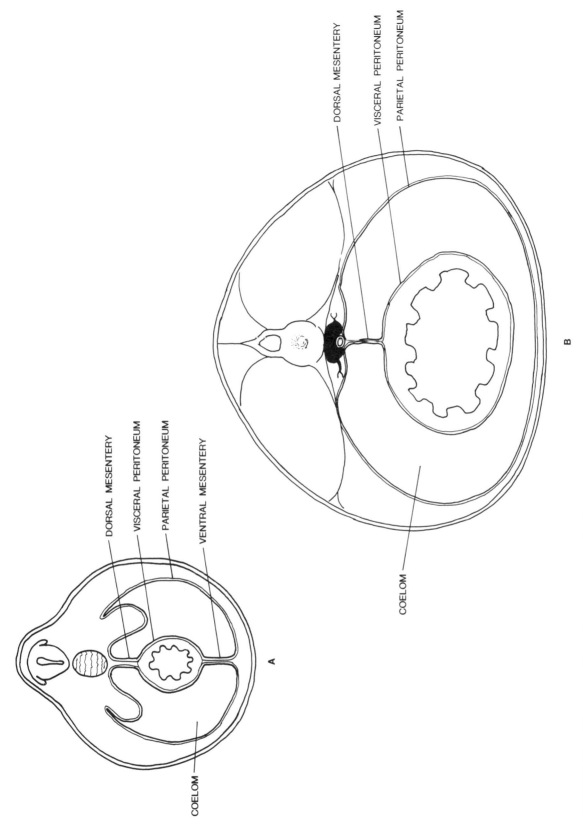

FIGURE 28. Diagrams of vertebrate serous (coelomic) membrane formation. *A* is an early embryonic stage with two elongated fluid-filled coelomic tubes and dorsal and ventral mesenteries. *B* is a typical adult stage after the disappearance of the ventral mesentery and the fusion of the two coelomic chambers.

COELOMIC MEMBRANES

The internal organs that undergo considerable movement, such as the heart and intestines, are enclosed by membranes and suspended by mesenteries within the fluid-filled coelomic cavities.

The coelomic cavity around the heart is the *pericardial cavity* and is continuous with the coelomic cavity surrounding the other visceral organs caudal to the heart. The passageway from the pericardial cavity to the body cavity is called the *pericardioperitoneal canal*. The body coelomic cavity is usually termed the *pleuroperitoneal* cavity but this is a misnomer for the shark. The term, "pleura" is used anatomically in reference to lungs and is derived from the Greek word for "rib" or "side." The shark, of course, has no lungs and furthermore lacks the homologous "air sacks" (or swim bladder) found in the bony fishes. Consequently, the only organs in the shark body cavity are peritoneal rather than pleuroperitoneal. Nevertheless the term "pleuroperitoneal" is applicable to bony fish, amphibians, and reptiles and is in general usage for this region in the shark.

Embryonically, the coelomic cavities develop laterally paired chambers extending from the heart to the rectum with both dorsal and ventral mesenteries. Figures 28 and 29 are diagrammatic cross sections of this arrangement. The membranes lining the body wall of the coelomic cavity are *parietal* membranes and those on the surface of the body organs are *visceral* membranes. Above and below the visceral organs the walls of the two coelomic sacs are in contact and form the double-walled *mesentery,* either *dorsal* or *ventral* depending on the location in reference to the visceral organ they attach to. The mesenteries of the heart are transitory during embryonic development and the adult heart has no mesenteries attached to it. In addition, a constriction develops between the pericardial cavity of the heart and the coelom of the intestinal region so the pericardial cavity becomes isolated in the adult, ventral to the pharynx and cranial to the body cavity. The membrane remaining on the surface of the heart is the *visceral pericardium* and that on the wall of the pericardial cavity is the *parietal pericardium.*

Most of the ventral mesentery of the body cavity also disappears during embryonic formation, but a portion called the *falciform ligament* remains at the extreme rostral end of the body cavity from the ventral body wall to the liver and the *gastrohepatic ligament* from the liver to the stomach. The ventral mesentery undergoes some unusual contortions during embryonic development so a portion of the mesentery loops upward, contacts, and fuses with an area of the dorsal mesentery. After contacting the dorsal mesentery, the ventral mesentery continues on to its insertion on the ventral border of the stomach, and the portion from the ventral body wall to the dorsal mesentery disappears. This small piece of ventral mesentery encloses the common bile duct, the hepatic portal vein, and the hepatic and pancreaticomesenteric arteries. That portion of the ventral mesentery to the common bile duct is the *hepatoduodenal* ligament and the portion from the common bile duct to the ventral surface of the stomach is the *gastrohepatic ligament*. The arrangement of some of these membranes in the adult shark are shown in figure 29. Note that the ventral surface of the stomach is not in a ventral position in figure 29 but is actually pointed to the right side of the shark. Note also that the attachment of the dorsal mesentery to the stomach is directed toward the left. This position of the stomach is due to a rotation of this portion of the gut tube during embryonic formation. The dorsal mesentery attachment on the remainder of the intestine is located dorsally as might be expected. As mentioned earlier, the ventral mesentery is lacking on the rest of the intestine.

The dorsal mesentery is named according to the particular part of the intestinal tract it attaches to. That portion attaching to the stomach is the *mesogaster,* the mesentery to the duodenum and rostral end of the spiral intestine is the *mesentery proper* or *mesointestine,* and the membrane attaching to the rectum and enclosing the rectal or digitiform gland is the *mesorectum.* Chondrichthyes are unique in the arrangement of the dorsal mesentery since in these fishes there is a gap in the dorsal mesentery between the mesentery proper and the mesorectum. Other gaps also occur between the spleen and the spiral intestine. These gaps are probably due to the embryonic formation of the spiral intestine, which undergoes considerable coiling during its formation. If the mesentery were to remain attached during this coiling, the membrane would wrap around and choke the spiral intestine. Thus the embryonic intestinal coiling prevents the association of cells of the embryonic membrane so the adult dorsal mesentery does not attach to the spiral intestinal wall, but the membrane does cover the blood vessels (rostral mesenteric and lienogastric arteries and dorsal intestinal vein) serving the spiral intestine.

The gonads (testes in the male and ovaries in the female) also have mesenteric attachments to the body wall but these mesenteries are on either side of the midline dorsal mesentery to the intestinal tract. In the male the mesentery to each testis is the *mesorchium.* In the female each ovary is suspended by a *mesovarium.* In the adult female the enlarged oviducts and uteri are suspended by a *mesotubarium* on either side of the midline.

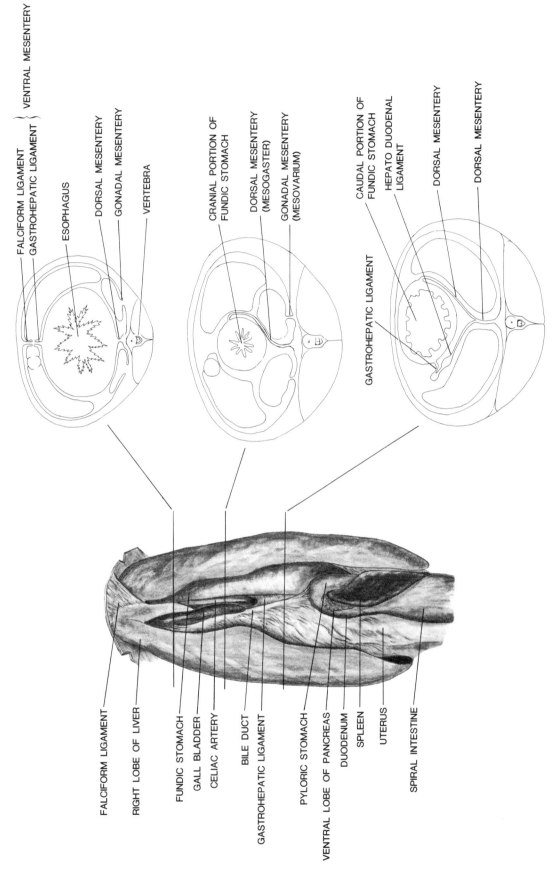

FIGURE 29. Ventral view of the shark visceral cavity and cross section diagrams at three different levels of the body cavity to illustrate the coelomic membranes.

Chapter 5
Oral and Pharyngeal Cavities

You have already opened the body cavity to examine the mesenteries and coelomic cavities. If your specimen is a female, do not disturb the falciform ligament during your examination of the digestive tract because that membrane will be important in your study of the reproductive system.

Pry open the mouth (your hemostatic forceps are a good tool for this purpose) and insert your scissors (blunt tip in the mouth) at the corners of the mouth, on the side you have previously skinned, and make a clean cut from the mouth through the articulation of the jaws and the mid portion of the gills, separating the epibranchials and ceratobranchials. This dissection will enable you to examine the mouth and pharynx.

MOUTH AND PHARYNX

1. The *mouth* is roughly defined as the cavity between a line just external to the teeth and the internal openings of the spiracles.
2. *Teeth* of sharks are homologous to the placoid scales that cover the body. Each tooth has an outer "cap" of *enamel** over a thick inner body of *dentine*. The teeth of all vertebrates are developed from the dermis of the skin and, when associated with bone, the teeth are associated with *dermal* not *cartilage* (endochondral) bone. As pointed out earlier (p. 9) the shark has a cartilaginous skeleton virtually devoid of bone. In the shark the dentine is based on individual plates of dermal bone that are securely anchored to the fibrous connective tissue of the dermis. The dermis separates the teeth from the deeper cartilaginous jaws.

   All the teeth are similar in shape and are replaced regularly as the functional tooth is worn down or broken. The replacement teeth lie in an epithelially lined cleft on the oral surface of the jaw (see fig. 30).

3. The *pharynx* is the cavity between the mouth and the esophagus. The *esophagus* is very short and may be recognized as the *papillae* lined opening to the stomach.

   The *gill slits* open on the lateral walls of the pharynx and are guarded by short *gill rakers* that may project slightly into the pharynx.

4. The *spiracle* marks the cranial border of the pharynx. Pass your blunt probe through the external opening of the spiracle until it emerges into the pharynx. Note the relationship of the spiracular passage to the jaws. The spiracle represents the first branchial slit and functions as a passageway for water into the pharynx, thus providing an accessory respiratory passage when the mouth is obstructed.

5. *Gill pouches,* the chambers between the internal and external gill slits, are lined with the gill filaments. Your earlier dissection opened the gill pouches on one side, so you may now examine these chambers and the internal and external gill slits that are still intact on the opposite side.

6. *Gills* are supported by the branchial arches (epibranchial and ceratobranchial cartilages) and the *gill rays* that extend laterally from the arches. This gill skeleton was examined earlier (p. 12). Extrabranchial cartilages lie at the tips of and at right angles to the gill rays but these are lost in skeletal preparations so you were not able to observe these cartilages earlier (see fig. 31). Afferent and efferent branchial arteries parallel the branchial arches and branches of these arteries serve the gill filaments (see Circulatory System, pp. 43–46). The gill filaments are arranged as primary lamellae paralleling the gill rays. Note the set of lamellae on the cranial wall of the gill pouch and another set on the caudal wall. These lamellar sets are *hemibranchs* of two separate gills. Two hemibranchs on the opposite sides of the same gill arch (branchial arch) make up a *holobranch*.

---

*It was originally believed (Owens, 1845) that the outer covering of the shark tooth was dentin that developed more densely than in subsequently formed layers. This conclusion was based principally on histological (see review by Salomon, 1969) and embryological (Kvam, 1953) studies. More recent biochemical and electron microscope studies indicate that a thin outer layer of the "cap" is true enamel (Moss, 1969; Shellis and Miles, 1974) and the inner layers of the tooth "cap" are produced by both enamel and dentin forming cells (Moss, 1977).

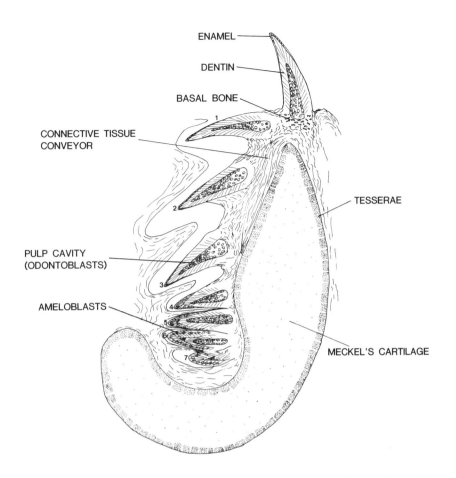

FIGURE 30. Cross section through the lower jaw (Meckel's cartilage) of a shark (*Negaprion brevisostris*) with the superficial tissue removed. The functional tooth is erect at the crest of the jaw and the replacement teeth are at right angles to the functional tooth. The connective tissue layer acts as a conveyer belt to pull the replacement teeth into position. (With permission from photographs by P. J. Boyne, 1970, *J. Dent. Res.* 49 (3): 556–60.)

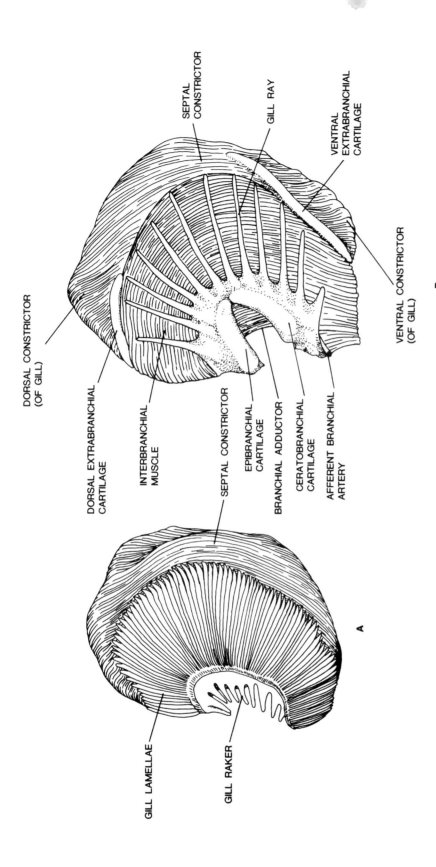

FIGURE 31. Surface features of an opened gill arch (*A*) and a gill arch with the filaments removed (*B*) to show the gill rays and interbranchial muscles.

Gill rakers extend from the base of the gill arch and prevent large particles from entering the gill pouch. The gill filaments of the spiracular pouch are poorly developed and not capable of respiratory gas exchange. These nonfunctional gill filaments are called *pseudobranchs*.

SUGGESTED READINGS

Baumgarten-Schumann, D., and J. Piiper. 1968. Gas exchange in the gills of resting unanesthetized dogfish (*Scyliorhinus stellaris*). *Respiration Physiology* 5:317–25.

Boylan, J. W. 1967. Gill permeability in *Squalus acanthias*. In P. W. Gilbert, R. F. Mathewson, and D. P. Rall, eds., *Sharks, skates, and rays*. pp. 197–206. Baltimore: Johns Hopkins Press.

Crespo, S. 1982. Surface morphology of the dogfish *Scyliorhinus canicula* gill epithelium and surface morphological changes following treatment with zinc sulfate: A scanning electron microscope study. *Marine Biology (Berlin)* 67 (2): 159–66.

Kvam, T. 1953. On the development of dentin in fish. 1. *Squalus acanthias* Linnaeus. *J. Dent. Research* 32:280–86.

Moss, M. L. 1969. Phylogeny and comparative anatomy of oral ectodermal-ectomesenchymal inductive interactions. *J. Dent. Res.* 48:732–37.

Olson, K. R., and B. Kent. 1980. The microvasculature of the elasmobranch gill. *Cell Tissue Res.* 209:49–63.

Salomon, C. D. 1969. Dentin of *Carcharhinus milberti* (Shark): A comparative histological and histochemical study. *J. Dent. Res.* 48:196–205.

Satchell, G. H. 1959. Respiratory reflexes in the dogfish. *J. Exper. Biol.* 36 (1): 62–71.

Satchell, G. H. 1968. A neurological basis for the coordination of swimming with respiration in fish. *Comp. Biochem. Physiol.* 27: 835–41.

Shelton, G. 1970. The regulation of breathing. In *Fish Physiology,* ed. W. S. Hoar and D. J. Randall, vol. IV, pp. 310–14. New York: Academic Press.

Shellis, R. P., and A. E. W. Miles. 1974. Autoradiographic study of the formation of enameloid and dentine matrices in teleost fishes using tritiated amino acids. *Proc. R. Soc. Lond. (Biol.)* 185: 51–72.

Chapter 6
Digestive System

THE FEEDING MECHANISM

The hyomandibular cartilages of the spiny dogfish extend at right angles from the cranium to the mandibular arch and thus the jaws do not protrude forward during feeding as they do in those sharks with sharply angled hyomandibula. In addition, the preorbital processes of the palatoquadrate of *Squalus* are very long and extend to the top of the orbit. These features allow the jaws to rotate and protract downward and at the same time resist the forces applied to the lateral walls of the mouth during feeding.

The solidly based jaw components coupled with broad, sharp cutting teeth provide these sharks with an efficient cutting mechanism. Thus the spiny dogfish either takes bites out of larger prey or severs into pieces the smaller fish or squid that make up the greater part of their diet.

Other sharks feed by protruding their jaws and biting a large chunk with the protruded jaws. This gouging type of feeding is performed on prey that is too large to be surrounded by the shark's open jaws.

A third feeding mechanism is found in still other elasmobranchs. These forms crush their prey and engulf the crushed prey with a sucking action of the jaws.

The hyomandibula is an important unit of the jaws of all jawed fishes (the Agnatha lack jaws). The upper and lower jaws have transformed from branchial arches. An overly simplified explanation of this origin is that the palatoquadrate is derived from the epibranchial cartilage and the lower jaw (Meckel's cartilage) corresponds to the ceratohyal. The major arguments over this description concern the specific branchial arches that were involved in jaw formation.

In any case, the primitive jaws serve as grasping organs and are not self-supporting. The hyoid arch, and especially the hyomandibula of the hyoid arch, serves to support the grasping jaws. Subsequent modification of the hyomandibular-jaw relationship allows several modifications of the fish's feeding mechanisms and, consequently, behavior.

The Alimentary Tract (figs. 32, 33)

1. The *esophagus* is a constricted region between the pharynx and stomach lined with elongate papillae. Unlike the esophagus of tetrapod vertebrates with a flexible neck, that of the shark is very short and wide. You may see the esophagus more clearly by continuing the cut you previously made through the jaws (see p. 32, chapter 5) through the side of the esophagus and stomach.

2. The *stomach* is J-shaped. The long straight part is the *fundic portion,* and lies nearest the heart. The short "limb" beyond the bend is the *pyloric portion*. The outside of the bend is the *greater curvature* and the inside is the *lesser curvature*. The stomach ends at the *pylorus* or passageway between the stomach and duodenum. Here you should feel the thickened involuntary muscular ring, which is the "gate" that closes or opens the pylorus. This muscular "gate" is the *pyloric sphincter*.

   Slit open the stomach and remove any food it may contain (such as squid, octopi, or small fish). These, if present, may show signs of having been partly digested. The wall of the stomach is composed chiefly of muscle (derived from mesoderm) covered externally by visceral (splanchnic) peritoneum and internally by a mucous membrane, the *mucosa*. The latter consists mostly of the lining epithelium of the gut (the only part of the wall derived from *endoderm*) and an underlying layer of fibrous connective tissue, the *submucosa*. The epithelium, besides lining the stomach, forms the microscopic gastric glands. The mesentery of the stomach is termed the *mesogaster*. Now wash out the stomach, and inspect the *rugae* (folds) on its inner lining. Compare the rugae with the papillae seen cranially as the lining of the esophagus. Continue the slit through the pylorus in order to note the size and form of the *pyloric sphincter*.

3. The *duodenum* extends only from the pyloric sphincter to the valvular intestine. Its mesentery is the *mesoduodenum*.

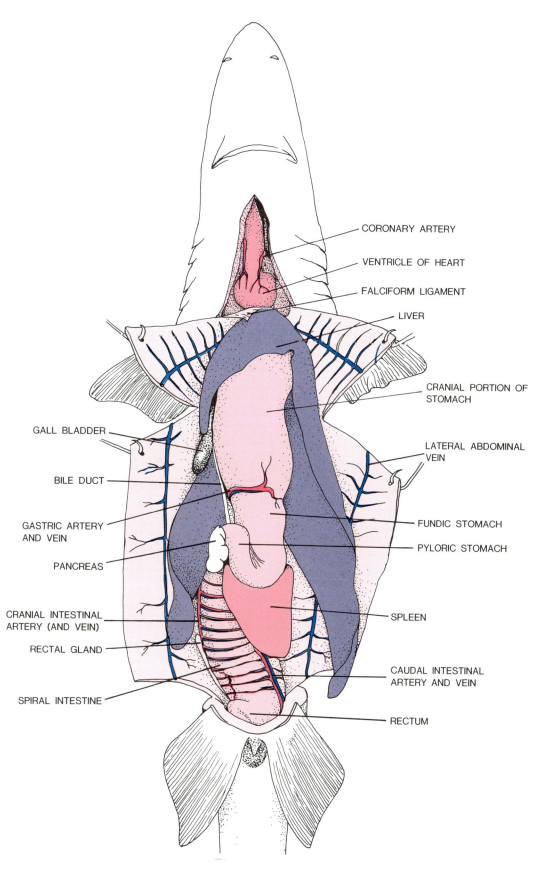

FIGURE 32. Ventral view of the shark viscera and heart.

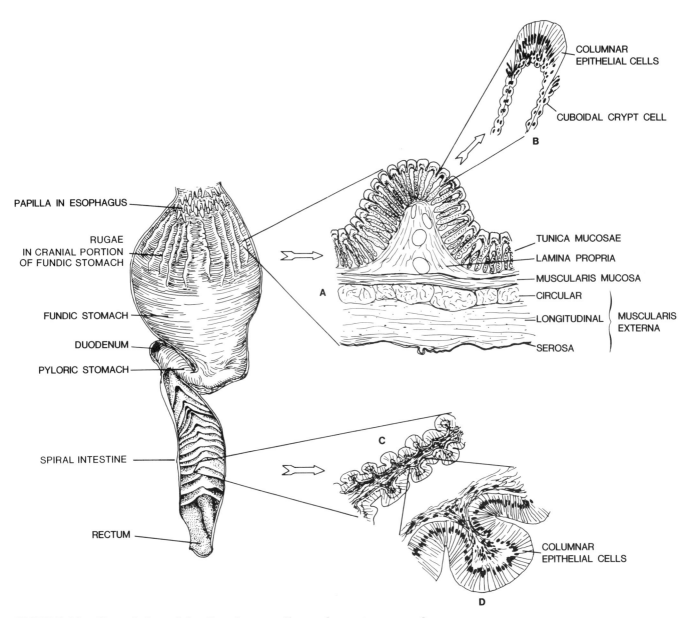

FIGURE 33. Ventral view of the digestive tract dissected open to expose the mucosal surface of the tract and views of light microscope sections of the mucosa of the stomach and typhlosole. *A* is a section through a single rugae of the stomach (50 ×). *B* is an enlargement (100 ×) of the epithelium and crypt. *C* is a section through a single "leaf" of the typhlosole and *D* is an enlargement (100 ×) of the typhlosole epithelium.

4. The *ventral pancreas* is a flattened white gland in the cranial curve of the duodenum. Its "mate," the *dorsal pancreas,* is a long, slender white gland reaching from the ventral pancreas along the dorsal side of the stomach and duodenum to the triangular *spleen* at the curve of the stomach. The narrow band of pancreas connecting its dorsal and ventral lobes is the *isthmus.* In addition to secreting essential digestive enzymes, the pancreas contains insulin secreting *isles of Langerhans,* which are visible only in microscopic sections. The *pancreatic duct,* from the ventral lobe to the right side of the duodenum, is too slender to be readily seen by dissection.
5. The *liver* (see figs. 32 and 42) is the largest organ in the pleuroperitoneal cavity, consisting of *right and left lateral lobes* and a ventral *median (cystic) lobe* with the *gall bladder* on the right border of the median lobe.
6. The *bile duct* collects bile from the liver into numerous branches leaving the liver as the *hepatic duct* (Latin *hepar* =, liver), which sends a branch, the *cystic duct* (Greek *kystis* =, bladder), to the *gall bladder.* The main bile duct continues to the duodenum where it empties the bile. The bile is an alkalinizing fluid that aids, indirectly, in the process of enzymatic digestion. Some dissection may be necessary in order to trace the bile duct from the cranial end of the duodenum to the liver by following the *hepatoduodenal ligament.*
7. The *gall bladder,* a greenish sac in the median lobe of the liver, is at the end of the short side branch (cystic duct) of the bile duct. The gall bladder is the thin-walled reservoir for the bile.
8. The *spiral intestine* is the broad, short, intestine extending caudally from the duodenum. Cut the ventral wall of the spiral intestine to observe the structure of the *spiral valve* (fig. 33). The spiral fold is also known as a *typhlosole,* a much more elaborate structure than the familiar typhlosole of the earthworm. In sharks, it is developed to such a height that the lumen of the intestine is divided to form a spiral passage that greatly increases the surface area for intestinal absorption. The same result is accomplished in higher vertebrates by a long, coiled intestine lined with circular folds and studded with numerous fingerlike microscopic villi.
9. The *rectum* is the short terminal portion of the gut between the valvular intestine and the cloaca.
10. *Microscopic structure*—The walls of the stomach, duodenum, and rectum are similar in that each consists of four layers (coats). These are, from within outward, the *mucosa, submucosa, muscularis,* and *adventitia.* The basic structures of the duodenal wall viewed in cross section include the following details (also see fig. 33).
    a. The *mucosa* consists of three layers that are, from within outward, a one layered columnar epithelium with mucus secreting goblet cells; a delicate connective tissue layer; and the *lamina propria,* which includes connective tissue cells, blood and lymph capillaries, and nerve fibrils.
    b. A very thin *muscularis mucosae* of fine smooth muscle fibers separates the *submucosa* that consists of a few scattered fibroblasts, blood vessels, and fine nerve fibers from the lamina propria.
    c. The outermost layer is the *muscularis externa* with a thick inner circular and a thinner outer longitudinal layer of smooth muscle fibers, surrounded by an outer thin *serosa* of visceral *pleuroperitoneum.*

    Sometimes small patches of lymphocytes are present in the submucosa and tiny, stellate nerve ganglia may be found between the muscularis and the submucosa or just beneath the submucosa and next to the muscularis mucosae.

NOTE: Conspicuous organs not belonging to either digestive or respiratory systems that may be noted at this time include the *spleen,* a dark triangular body attached to the greater curvature of the stomach. It is a glandular organ concerned with the maintenance of the blood. The *gonads,* ovaries in the female and testes in the male, are large paired ovoid structures attached to the wall of the body cavity dorsal to the stomach and liver.

The *kidneys* are a pair of brownish, ribbon shaped organs lying close together on either side of the dorsal midline where they extend most of the length of the pleuroperitoneal cavity. Gonads and kidneys will be described in greater detail in chapters 8 (Excretory System) and 9 (Reproductive Systems).

SUGGESTED READINGS

Moss, S. A. 1977. Feeding mechanisms in sharks. *Amer. Zool.* 17 (2): 355–64.

Chapter 7
Circulatory System

The major emphasis in this chapter is devoted to the vessels that the blood flows through. However, some brief remarks about the blood itself may be useful.

Blood consists of a liquid *plasma* and suspended cells. The plasma of elasmobranchs has an osmolarity very similar to the seawater environment. The high osmolarity is maintained by the high concentration of urea and trimethylamine oxide that are retained by these fishes. With a few exceptions the high concentration of urea in plasma is toxic in bony fish, amphibians, and higher vertebrates. It is probable that the insensitivity to uremic (high urea concentration) poisoning is a primitive condition allowing these animals to survive in a wide variety of habitats from marine (salt) to fresh waters.

The blood cells of elasmobranchs, as with other vertebrates, are grouped as red cells or erythrocytes and as white cells or leukocytes and thrombocytes. All cells, red, white, or thrombocyte, are nucleated and although they are not as large as the cells of amphibians and some bony fishes, they are larger than the blood cells of reptiles, birds, and mammals. The larger size of red blood cells decreases the total amount of hemoglobin that can contact and bind with respiratory gases (oxygen and carbon dioxide). Consequently, the binding capacity of shark blood is much less than that of mammals unless the mammalian blood is diluted to the same hemoglobin level as the blood of sharks.

In addition to transporting nutrients, gases and wastes, the blood also transports *temperature*. Blood is warmed by the heat generated by muscle contraction (and by high environmental temperature) and cooled by contact with low environmental temperature. Cooling is especially critical in the gill capillaries where not only gases are exchanged but heat is also lost to the cold sea water. Some large sharks (and bony fishes) have compensated for this heat loss by the development of an intricate plexus of blood vessels (termed a *rete mirable*) in the dark muscle along the side of the shark trunk.

DISSECTION INSTRUCTIONS

Most of the circulatory system is already exposed, or nearly so. Stripping away the epithelium lining the roof of the oral cavity and pharynx will expose the *efferent branchial aortic arches*. Lay the shark on its back and open the flap of the lower jaw as in your previous dissections (Chapters 4, 5, and 6; figs. 27 and 32). Grasp the mucosal lining at its cut edge near the caudal end of the pharynx with hemostatic forceps and strip it away by pulling it toward the mouth. Repeat this process until all the dorsal aortic arches are exposed. In a similar manner strip away the mucosa lining the lateral wall of the pharynx (the gill arches) and the lining of the gill slits. The pericardial cavity should have been exposed with your earlier dissection (see fig. 32). If this has not been done, review figures 27 and 32 and cut through the coracoid cartilage and hypobranchial muscles on both sides (right and left) of the liver forward to the inner border of the lower jaw. Make a transverse cut through the transverse membrane and ventral to the liver. Do not cut into the liver because this will destroy the hepatic vein that opens into the sinus venosus from the liver. When you finish the three cuts remove the pyramid-shaped segment covering the pericardial cavity and heart. This dissection will appear similar to figure 32. If additional dissection is necessary, use a blunt probe to loosen the tissue and dissect very carefully to avoid cutting the aorta or afferent branchial arches (figs. 36 and 38).

THE PERICARDIAL MEMBRANES

The tough gray-white membrane lining the pericardial cavity is the *parietal pericardium*. As mentioned earlier (p. 30) the pericardial cavity is continuous with the pleuroperitoneal cavity through the *pericardio-peritoneal canal*. It will be difficult to locate this canal without removing most of the sinus venosus in the pericardial cavity and liver in the pleuroperitoneal cavity. Remember to come back to this canal after you have completed your dissection of the heart and then use a blunt probe to locate it.

The *visceral pericardium* is on the surface of the heart. Although there are no mesenteries in the pericardial cavity, these mesenteries were briefly present during embryonic development and disappeared before hatching.

THE HEART

Your view of the heart (figs. 32 and 36) is of the ventral surface of the *ventricle* and *conus arteriosus*. The other chambers of the heart (*atrium* and *sinus venosus*) are dorsal to the ventricle and will require additional dissection in order to examine them.

The four chambers of the heart are arranged in a linear series and are described below from caudal to cranial. Spread open the pericardial cavity and identify these chambers externally (the ventricle and conus arteriosus will be seen first). Lift the caudal end of the ventricle in order to see the sinus venosus and atrium.

1. *Sinus venosus*, a thin-walled triangular sac with its base against the *transverse septum*. Blood is returned from the body through veins and sinuses and empties into this part of the heart (figs. 34 and 38).
2. *Atrium* is also thin-walled and bulges laterally forming a bilobed chamber that receives blood from the sinus venosus.
3. *Ventricle*, the ovoid, thick-walled muscular "pump" that lies in the caudoventral portion of the pericardial cavity.
4. *Conus arteriosus* is an elongated arterial cone with walls containing elastic tissue and heart muscle to assist the ventricle in forcing venous blood forward into the gills to be reoxygenated. It extends from the rostral end of the ventricle to the rostral margin of the pericardial cavity (fig. 34).

Now open the sinus venosus with a transverse incision from the left *common cardinal vein* to the incision passing from the mouth to the left side of the stomach. Next, open the left side of the atrium longitudinally; then cut the ventricle open in the frontal (horizontal) plane beginning caudally and continuing forward for about four-fifths of the length of the ventricle. Continue the incision along the right side of the conus arteriosus nearly to the first pair of afferent branchial arteries. Now wash out all of the blood from the heart. Identify and locate the following:

5. The paired *common cardinal veins* (ducts of Cuvier) enter the sinus venosus at each caudolateral angle near the junction of the common cardinals and the sinus venosus. Find the openings of the *caudal cardinal sinuses*.
6. The small *inferior jugular veins* open at the rostrolateral angle of the sinus venosus. The *hepatic sinuses* open near the center of the caudal wall of the sinus venosus.

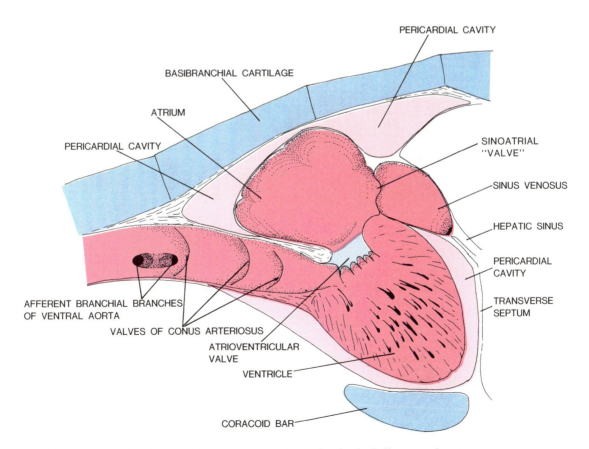

FIGURE 34. Midsagittal section of the shark heart *in situ* including attachments of the pericardial membrane.

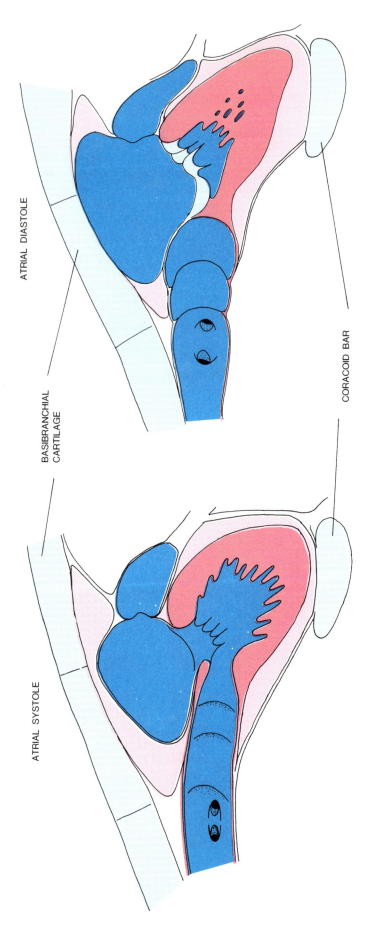

FIGURE 35. Diagrammatic sagittal views of the shark heart in (*A*) ventricular diastole (atrial systole) and in (*B*) ventricular systole (atrial diastole). The pericardial membrane is attached to the coracoid bar ventrally and the basibranchial cartilage dorsally. These attachments do not allow the pericardial cavity (light red) to constrict with ventricular systole thus creating a potentially negative pressure in the pericardial cavity that "sucks" blood (blue) into the atrium from the sinus venosus and cardinal sinuses.

7. The *sinoatrial valve* guards the opening from the sinus venosus into the atrium.
8. The *atrioventricular orifice* from the atrium to the left side of the ventricle is closed by the *atrioventricular valve*.
9. The *semilunar valves* are 9 to 12 cuplike flaps in sets of three (or four in some large sharks) cups per set. Each set of three cups encircles the conus arteriosus; usually two groups lie at the caudal end, and one set, with much larger flaps, lies at the rostral end (fig. 34). Insert a blunt probe into one of the valve flaps. The edges of the cups meet in the center of the lumen of the conus arteriosus each time the ventricle expands to draw blood from the sinus venosus and atrium. This prevents back flow of blood from the conus into the ventricle.

THE VENTRAL AORTA AND AFFERENT BRANCHIAL ARTERIES

The conus arteriosus leads cephalad from the pericardial cavity to the ventral aorta (fig. 36). Follow this vessel forward by carefully removing hypobranchial muscles and connective tissue until you reach the lower jaw (Meckel's cartilage).

1. The *ventral aorta* is the rostral continuation of the conus arteriosus that eventually bifurcates to form two widely arching vessels just caudal to the lower jaw. Identify, but do not destroy, the small *coronary arteries* along the conus and ventricular walls.
2. *Afferent branchial arteries* originate as three paired lateral branches plus the final bifurcation of the ventral aorta. The arteries formed by bifurcation of the aorta divide again so a total of five pairs of afferent branchials carry blood to the gills. Carefully trace these vessels on the side of your dissection with the uncut gill arches. The most caudal afferent enters the interbranchial septum of the sixth branchial bar. Beginning cranially, the first afferent branchials are distributed to the interbranchial septum of the second branchial bar, the second pair branch to the septa of the third bar, the third branch to that of the fourth, and the fourth to that of the fifth branchial bar. The fifth is the most caudal afferent vessel and it enters the sixth interbranchial septum. Follow the third or fourth afferent branchial of the left side to the interbranchial septum, and slit open the septum to expose the tiny filamental arteries that branch to the gill lamellae.

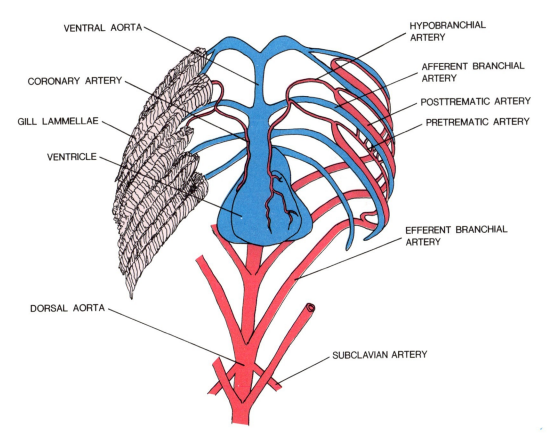

FIGURE 36. Ventral view of the heart and aortic arches of the shark.

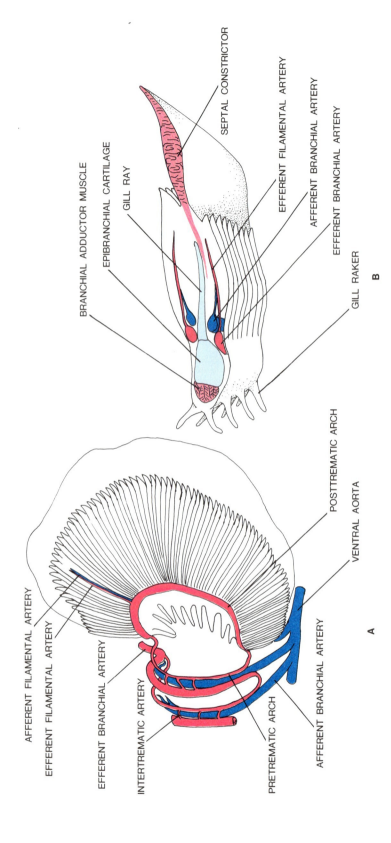

FIGURE 37. Major blood vessels of the gill arches (*A*) and cross section (*B*) of a gill arch. Note: *B* is a section through *A* and turned 90 degrees to show the cut surface.

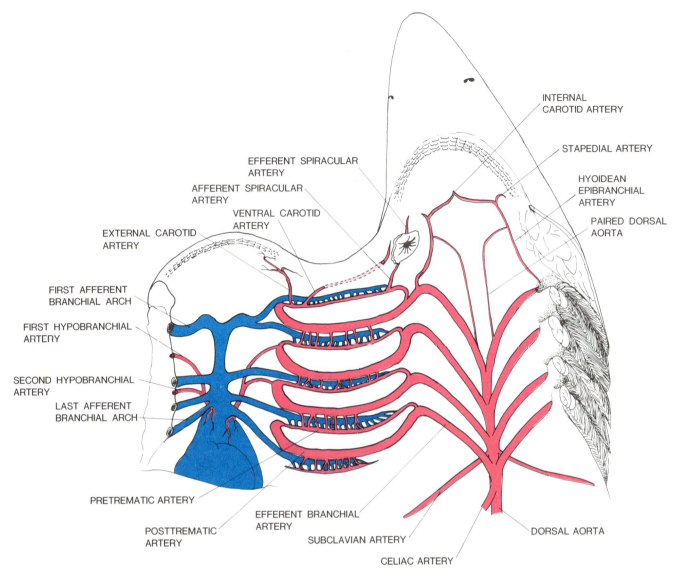

FIGURE 38. Dorsal view of the heart and ventral aorta and ventral view of the major vessels of the branchial arches.

GILL CIRCULATION

Circulation through the shark gill is an extremely complex excursion through microscopic vessels and sinuses (that you will not be able to see in a gross dissection). Blood is delivered to the gill filaments by afferent filamental arteries that have branched from an afferent branchial artery. Blood from the afferent filamental artery goes to either the *lamellar sinus* or the *interlamellar sinus* where gas exchange between the blood and water bathing the gills occurs. Blood in the interlamellar sinus drains into venous networks and is returned to the heart by venous passageways. Lamellar sinus blood drains into an efferent filamental artery (which is the smallest vessel you will see by gross dissection; fig. 37). The *efferent filamental arteries* empty into an *efferent collector loop* that encircles each gill slit at the base of the filaments. The portion of the efferent collector loop on the rostral border of the gill slit is termed a *pretrematic* artery and that portion on the caudal border of the slit is the *posttrematic* artery. The posttrematic is much thicker than the pretrematic and is continuous, on the roof of the mouth, with an *efferent branchial artery*.

Each pretrematic is connected with the posttrematic of the next rostral collector loop by several *cross anastomosing arteries*. In addition to receiving oxygenated blood from the efferent filamental arteries and draining into the efferent branchial arteries, each collector loop has vessels that are unique to that particular loop as follows:

1. The first (most rostral) pretrematic artery gives rise to an *external carotid* artery, which serves the lower jaw, and a *ventral carotid* to the upper jaw and rostrum. The dorsal portion of the first

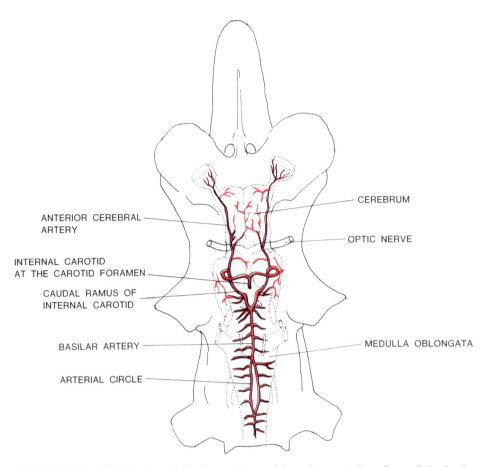

FIGURE 39. Distribution of the internal carotid on the ventral surface of the brain.

pretrematic has an *efferent spiracular* branch serving the pseudobranch of the spiracle. An *efferent spiracular* artery from the pseudobranch capillaries gives off a small branch to the upper jaw and snout and then enters the chondrocranium ventral to the stapedial artery and joins the internal carotid artery (fig. 39).

2. *Hypobranchial* or *commissural* arteries form at the ventral juncture of the pre- and posttrematic arteries of the second and third efferent loops. The two hypobranchials on each side join to form the *coronary* artery to the heart musculature.

3. The fourth collector loop receives blood from the isolated pretrematic vessel of the fifth gill slit via the cross anastomosing arteries to the posttrematic of the fourth loop.

EFFERENT BRANCHIAL ARTERIES

1. The four *efferent branchial arteries* are the dorsal portions of the aortic arches. These conspicuous paired vessels pass obliquely caudal from the dorsal angles of the gill slits. Remove all connective tissue from these arteries on the left side, tracing them until they disappear dorsal to the gill cartilages. Remove these cartilages by cutting across them and pulling the cut ends out toward the middorsal line.

The hyoidean epibranchial and afferent spiracular arteries branch from the first efferent branchial arch near its origin at the collector loop, and a rostral paired dorsal aorta branches just before the union of the first pair of efferent branchials in the midline.

2. The *hyoidean epibranchial* passes forward on the roof of the mouth and joins the paired dorsal aorta on each side at the level of the spiracle. The vessel is still termed the hyoidean epibranchial for a short distance forward until it divides into a medial *internal carotid* and a lateral *stapedial* artery.

3. The *stapedial* artery passes into the cartilaginous skull over the spiracular artery (remember you are looking at the ventral surface) and continues forward to the snout and orbit.

4. The *internal carotid* turns toward the midline where it joins its fellow from the opposite side and enters the carotid foramen to serve the brain, internal ear, and olfactory organs (fig. 39).

5. The paired *dorsal aortae* are slender anastomosing arteries between the first epibranchial arteries and the hyoidean epibranchials.

THE DORSAL AORTA

The *dorsal aorta* is formed by the union of the four main pairs of efferent branchial arteries. This large arterial trunk penetrates the transverse septum and enters the pleuroperitoneal cavity. From here the aorta has numerous branches to the viscera and muscles of the trunk and tail. The first branches are the *subclavian arteries*, which arise before the fourth pair of efferent branchial arteries join the dorsal aorta. All other branches are in the trunk or tail.

AND ITS BRANCHES

1. The *subclavian artery* supplies blood to the pectoral fin and adjacent body areas. The vessels originate from the aorta between the third and fourth efferent branchial arteries. Separate the esophagus from the body wall on the left side to expose the subclavians, and then trace the left subclavian dorsal to the large saclike *caudal cardinal sinus* and into the pectoral fin. Be careful not to damage either of the two arteries that branch from the subclavian.

    The following vessels are deep to the caudal cardinal sinus, the lateral abdominal vein, and to the lateral musculature so you may wish to wait until you have completed the venous dissection before observing these vessels.
    a. The *lateral artery* is a small vessel arising from the subclavian just lateral to the caudal cardinal sinus and extending caudad in the body wall parallel to the lateral line and deep to the *lateral abdominal vein*.
    b. The *ventrolateral artery* branches from the subclavian near the entrance of the lateral abdominal vein to the common cardinal vein. The ventrolateral artery then extends caudad between the lateral and midventral lines and branching into *segmental arteries* to the myotomes. In the caudal pleuroperitoneal cavity it forms an anastomosis with arteries entering the pelvic fins.
    c. The subclavian continues into the fin as the *brachial artery* after giving off the lateral and ventrolateral branches.
2. The *coeliac artery* is the first unpaired branch of the dorsal aorta as well as the first vessel to branch after the last (fourth) pair of efferent branchial arteries have joined the aorta. The coeliac branches from the ventral surface of the aorta just within the pleuroperitoneal cavity and supplies branches to the esophagus, stomach, liver, pancreas, and gonads. The esophagus, fundic end of the stomach, and gonads each receive small branches from the coeliac near its origin and before the main trunk continues caudally to enter the gastrohepatic ligament and trifurcate into the following branches.
    a. The *genital arteries* are paired branches from the coeliac (*ovarian arteries* in the female and *spermatic arteries* in the male). Genital arteries are poorly developed in immature sharks and are not usually injected with latex.
    b. The *gastric artery* supplies the stomach after dividing into *dorsal* and *ventral gastric arteries* (fig. 41B).
    c. The *hepatic artery* parallels the bile duct, passing forward to enter the substance of the liver.
    d. The *pancreaticomesenteric artery* passes dorsal to the pyloric end of the stomach supplying small branches to the stomach wall and ventral pancreas before branching into the following:
        i. The *duodenal artery* is a medium-size branch to the duodenum.
        ii. The large *cranial intestinal artery* passes over the right side of the valvular intestine to supply the folds of the spiral valve.
3. The unpaired *gastrosplenic* (lienogastric) *artery* arises from the ventral surface of the aorta near the middle of the pleuroperitoneal cavity and passes through dorsal mesentery to supply the spleen and greater curvature of the stomach.
4. The *cranial mesenteric artery* branches from the aorta just cranial to the gastrosplenic and supplies the left side of the valvular intestine by small branches to the valvular folds on the left side.
5. The *caudal mesenteric artery* branches from the aorta caudal to the gap in the dorsal mesentery and passes along the cranial margin of the mesorectum to the rectal gland.
6. *Renal arteries* may be seen by loosening the kidneys from the dorsal body wall and looking dorsal to the kidneys. Several of these leave the aorta to enter the kidney.
7. *Parietal arteries* are somatic branches from the aorta that pass along the myosepta to supply the body wall. One of the more cranial pairs of these vessels may become genital or *uterine arteries* in mature females.
8. The *iliac arteries* supply the pelvic fins. They arise from the aorta just cranial to the cloaca, make anastomotic connections with the caudal ends of the ventrolateral arteries, and send a network of branches into the cloacal walls before entering the pelvic fins.

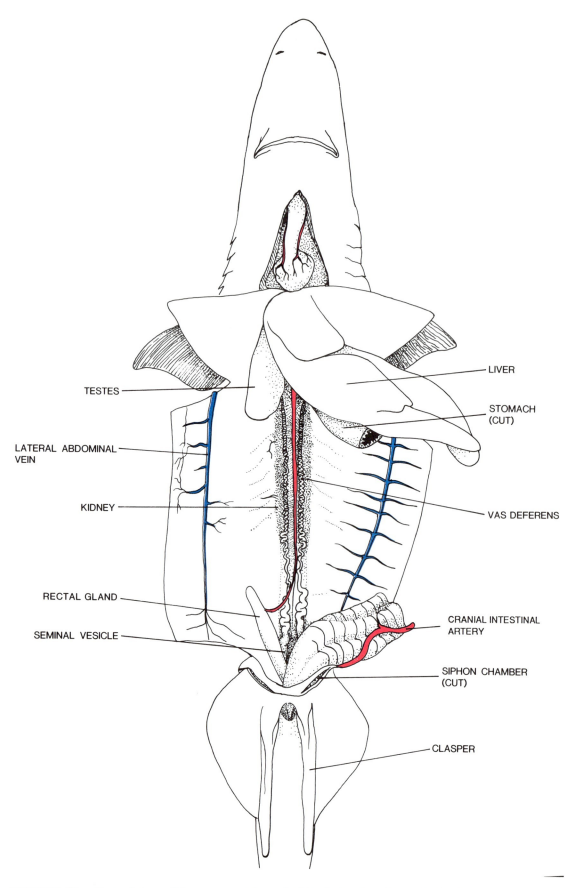

FIGURE 40. Ventral view of the aorta and lateral abdominal veins of a male shark.

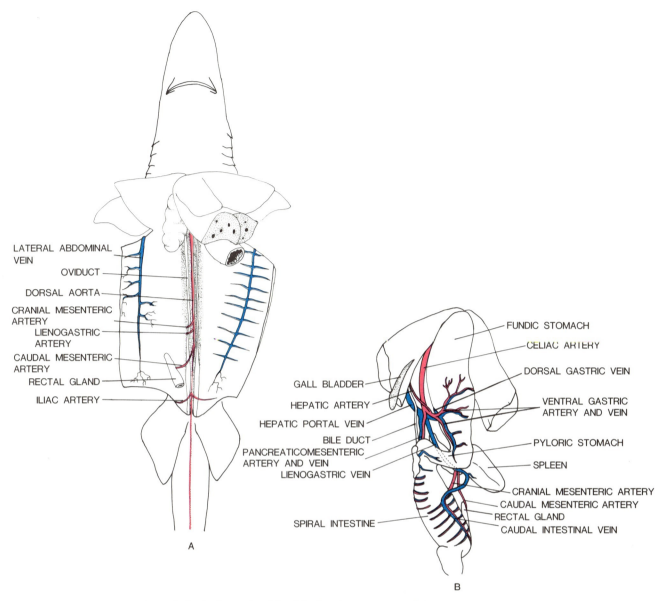

FIGURE 41. Ventral view of the body cavity (A) with the viscera removed to expose the dorsal aorta and its major branches. Figure B is a ventral view of the viscera with branches of the coeliac and mesenteric arteries and parts of the hepatic portal system.

9. The *caudal artery* is the unpaired continuation of the dorsal aorta to the end of the tail. It lies in the hemal canal just ventral to the vertebral centra and just dorsal to the caudal vein. The preserved shark is usually injected with red latex through this artery.

The coracobranchial muscles, previously noted (p. 25), may now be traced to their insertions on the ceratohyal and other branchial cartilages (figs. 13 and 25).

SYSTEMIC VEINS AND SINUSES

Systemic veins carry blood from tissues and organs directly back to the heart. Sinuses are more or less dilated venous channels such as the renal and cardinal sinuses, both of which return venous blood to the sinus venosus of the heart.

1. *Common cardinal veins* (ducts of Cuvier) are the large bilaterally placed tributaries to the sinus venosus, the left one has already been slit open. The right common cardinal vein should still be intact.

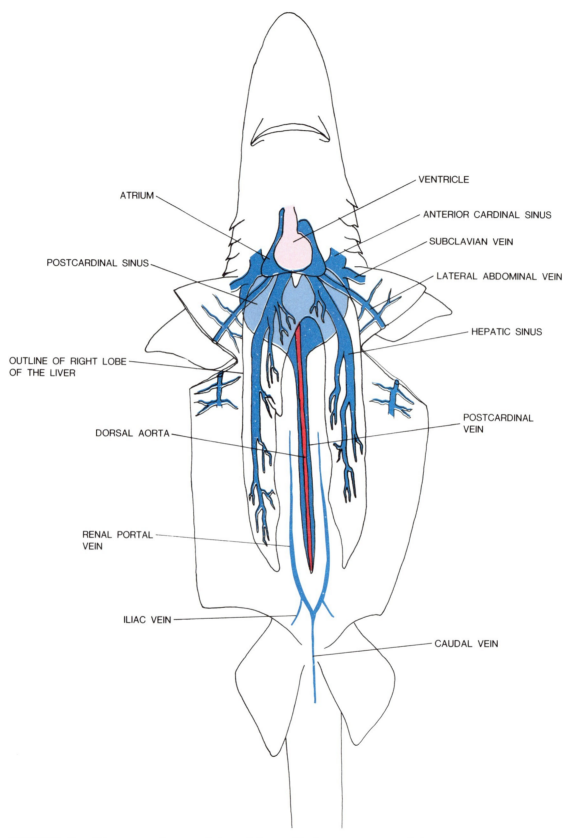

FIGURE 42. Diagrammatic ventral view of the cardinal venous system of the shark. The renal portal system may not be injected with colored latex and will therefore be difficult to find.

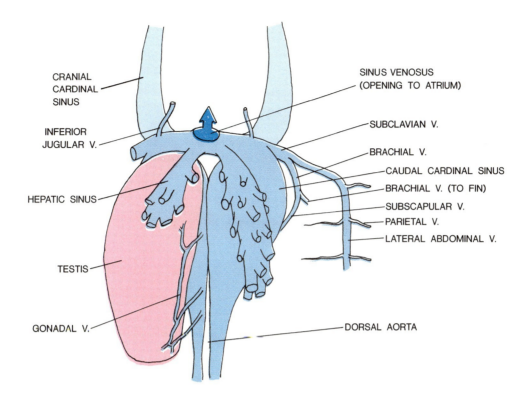

FIGURE 43. Ventral diagrammatic view of the cardinal venous system. All visceral organs except the right testis have been removed.

2. *Inferior jugular veins* bring blood from the floor of the mouth and ventral gill cavities to the sinus venosus. Probe the opening of one where it enters the sinus venosus, and follow the course of the vein as far cephalad as possible.
3. *Hepatic sinuses* can be found by first stretching the caudal wall of the sinus venosus to find a white fold at its center. On either side of this fold find an opening that extends through the coronary ligament into the hepatic sinus of the liver. Probe one of these and follow the probe with dissection of the liver so as to expose the sinus that gradually narrows to become the *hepatic vein.*
4. *Caudal cardinal sinuses* enter the sinus venosus just lateral to the openings of the hepatic sinuses. Find these openings and follow one of the sinuses with your probe as far caudally as possible. Move the end of your probe from side to side within the sinus and determine its width. Lift the viscera to one side in order to follow the probe. Having done this, reinsert your probe again about halfway back in the sinus, and continue probing to find its caudal limit. These two sinuses are joined across the midline at a point just cranial to the liver. Trace the connection with your probe.
5. *Genital sinuses* surround each gonad but may be difficult to see when devoid of blood or injection media. These sinuses are joined to the caudal cardinal sinuses (fig. 43).
6. *Parietal veins* are segmentally arranged, that is a pair arises from each myotome. They carry blood from the body wall into the caudal cardinal sinuses.
7. *Renal veins* are numerous small vessels that carry blood from each kidney into the adjacent caudal cardinal vein or sinus. Parietal and renal veins can only be seen when filled with blood or injection media.
8. *Cranial cardinal sinuses* return venous blood from the head to the sinus venosus. One of these was opened during the dissection of the brachial musculature, but if this dissection was omitted you may proceed by placing the shark on its ventral side and dissecting along the dorsal side of the left gills following the course of the lateral line. Deepen this incision until the large, smoothly lined cranial cardinal sinus is entered—it usually contains some darkened blood. Probe rostrally into this sinus and continue your incision along the probe rostrally above the spiracle and eye.
9. The *orbital sinus* should be partially exposed. Note that it completely surrounds the eye except for the corneal surface.
10. Next, the *interorbital sinus* may be followed from its connection between the two orbital sinuses by first finding an opening in the floor of the cranial cardinal sinus at the level of the caudal side of

the eyeball. Place your probe into this opening and pass it transversely through the interorbital sinus.

11. The *hyoidean sinus* leads from the floor of the cranial cardinal sinus at a point between the hyoid arch and the third gill arch passing ventrad along the lateral margin of the hyoid arch and into the inferior jugular vein, thus forming a vascular anastomosis between the two longitudinal vessels.
12. The *lateral abdominal* vein usually shows up conspicuously as it progresses cephalad along the pleuroperitoneal surface of the lateral abdominal wall. It receives parietal veins segmentally from the myotomes, as in the case of the cranial cardinal sinuses. The lateral abdominal veins empty cranially with the short subclavian veins into the common cardinals (ducts of Cuvier) just cranial to the expanded portions of the caudal cardinal sinuses (see figs. 42 and 43).
13. The *brachial* vein may be found by opening the lateral abdominal vein at its cranial end where the opening of the brachial joins the lateral abdominal. The brachial drains the veins of the pectoral fin, and its cut end may be seen where the pectoral girdle was transected.
14. The *subclavian vein* is a short continuation of the brachial from the pectoral girdle to the common cardinal.

    NOTE: Of course there is no *clavicle* in the shark and therefore no *subclavian* region, but this is the homologous vessel in the mammal receiving the brachial vein and delivering blood to the heart and this name is used in the absence of a standard terminology for fish anatomy.

15. The *cloacal vein* passes lateral to the cloacal opening and, as it progresses cranially, it joins the femoral vein.
16. The *femoral vein* empties into the lateral abdominal on the inner surface of the pelvic girdle. The opening of the femoral may be found by slitting the lateral abdominal vein near the middle of the base of the pelvic fin. Look for the femoral vein just beneath the skin on the dorsal side of the pelvic fin.
17. *Cutaneous veins,* ventral, dorsal, and lateral (paired) lie just beneath the skin. Segmental tributaries from the myotomes drain into the cutaneous veins. The cutaneous veins need not be traced in detail. Most of the above veins are illustrated in figures 32, 40, 41 and 42.

HEPATIC PORTAL SYSTEM

Portal systems, whether hepatic (in the liver) or renal (in the kidneys), consist of veins carrying blood from one capillary bed to another. Hepatic portal veins carry blood to the liver from the various parts of the digestive system. Veins carry blood *toward* the heart and arteries carry blood *away* from the heart. It is descriptively easier to trace both veins and arteries from the heart toward the tissues or visceral organs.

1. The *hepatic portal vein* parallels the bile duct in the gastrohepatoduodenal ligament (=lesser omentum). This vessel receives small veins from the bile duct but is formed mainly by three tributaries that converge and unite near the cranioventral tip of the dorsal lobe of the pancreas.
2. The *gastric vein* is the left tributary from the central part of the stomach where it is formed by the union of both the *dorsal* and *ventral cranial gastric veins,* each of which has cranial and medial roots.
3. The *gastrosplenic* (lienogastric) *vein* is the middle tributary of the hepatic portal. It traverses the dorsal surface of the duodenum and lies within the dorsal lobe of the pancreas from which it receives small *pancreatic veins.* A *caudal gastrosplenic vein* joins the gastrosplenic at the caudal end of the dorsal pancreas and brings blood from the spleen and adjacent regions of the stomach. The *caudal intestinal vein* also joins the gastrosplenic at the caudal end of the dorsal pancreas. This vein is formed by numerous parallel tributaries from the folds of the spiral valve of the intestine and a branch from the *rectal gland vein.*
4. The *gastrointestinal* (pancreaticomesenteric) *vein* is the right tributary of the hepatic portal vein. It is joined in the ventral pancreatic lobe by a small *pyloric vein* as it passes over the pylorus and by the *intraintestinal vein* from within the spiral valve. An important tributary to the gastrointestinal vein is the *cranial gastrosplenic vein.* It collects blood from the spleen and nearby pyloric stomach. A continuation of the gastrointestinal vein is the *cranial intestinal vein,* which passes along the right side of the valvular intestine from which it receives numerous parallel tributaries.

RENAL PORTAL SYSTEM

Make a fresh cut across the tail just caudal to the anus. Note again the haemal arch surrounding the caudal artery and vein. The caudal vein may be considered the origin of the renal portal system together with the iliac veins.

1. The *caudal vein* drains blood from the tail to a point just cranial to the anus where it bifurcates to join the paired iliac veins.

2. *Iliac veins* drain blood from the pelvic fins and join the bifurcated caudal vein to form the renal portal veins.
3. *Renal portal veins* provide blood to the kidney tubules (fig. 42). Probe forward with a flexible probe into the caudal vein and allow the probe to enter either fork of the renal portal system. Find the cranial end of your probe by carefully separating the long, brownish kidney along its lateral border from the dorsal body wall. Move your probe to see where it has reached into a renal portal vessel.
4. *Afferent renal veins* supply venous blood to the kidney and terminate in a capillary bed around each kidney tubule. Branches of the renal portal vein do not serve the kidney filtering system. The capillary knot in the renal corpuscle (glomerulus) is an organ of the renal artery (see fig. 45).

SUGGESTED READINGS

Carey, F. G. 1973. Fishes with warm bodies. *Sci. Am.* 228:36–44.

Daniel, J. F., and L. H. Bennett. 1931. Veins of the roof of the buccopharyngeal cavity of *Squalus suckleyi*. *Univ. of Calif. Pub. Zool.* 37:35–40.

Dzhumaliev, M. K. 1980. Capillary blood supply to the esophagus and the cardial section of the stomach in fishes of certain orders. *Vopr. Ikhtiol.* 20 (1): 79–85. (In Russian; English summary).

Gillian, L. A. 1967. A comparative study of the extrinsic and intrinsic arterial blood supply to brains of submammalian vertebrates. *J. Comp. Neurol.* 130 (3): 175–96.

Johansen, K., D. L. Franklin, and R. L. Van Citters. 1966. Aortic blood flow in free-swimming elasmobranchs. *Comp. Biochem. Physiol.* 19: 151–60.

O'Donoghue, C. H., and E. B. Abbot. 1928. The blood vascular system of the spiny dogfish, *Squalus acanthias* and *Squalus sucklii*. *Trans. R. Soc. Edinburgh* 55, part 3:326–890. (This is a classic upon which all subsequent descriptions are based).

Olson, K. R. and B. Kent. 1980. The microvasculature of the elasmobranch gill. *Cell Tissue Res.* 209 (1): 49–64.

Randall, D. J. 1970. The circulatory system. In *Fish physiology*, ed. W. S. Hoar and D. J. Randall, chapter 4, vol. IV. New York: Academic Press.

Satchell, G. H. 1970. A functional appraisal of the fish heart. *Fed. Proc.* 29:1120–23.

Chapter 8
Excretory System

Urea is formed following the breakdown of amino acids to ammonia. Ammonia is toxic and its conversion to urea or trimethylamine oxide is necessary to allow the concentration of nitrogenous wastes before excretion. The spiny dogfish retains a high level of urea and trimethylamine oxide in the blood as an osmoregulatory mechanism. Trimethylamine oxide is formed in some fishes but the spiny dogfish obtains this compound only in its food.

These waste compounds (urea and trimethylamine oxide) are excreted by the shark primarily through the kidneys but the rectal gland, the gills, and, of course, the skin are also capable of excreting waste substances.

The skin and gills are considered elsewhere and only the kidneys and rectal gland, whose primary function involves excretion, are considered here.

Locate and examine the following:

1. The *kidneys* are long, narrow strips of tissue (quite unlike the "bean-shaped" kidneys of mammals) on either side of the dorsal midline of the pleuroperitoneal cavity. The kidneys extend caudally as far as the cloaca but they do not reach the gonads at the cranial end of the pleuroperitoneal cavity in either sex. These structures are dorsal to the pleuroperitoneal membranes so they are "retroperitoneal" (retro = behind).

    Separate the kidney from the body wall along its lateral border and notice the varying thicknesses of the kidney at different points. The cranial, thinner portion is absent in females and the thick caudal portion is functional. A tough, glistening *caudal ligament* lies between the two kidneys dorsal to the renal portal veins and dorsal aorta.

    The shark kidney is an *opisthonephros* as distinguished from a *holonephros* (a serially segmented kidney found in larval Agnathans) and a metanephros (a caudally compacted kidney found in reptiles, birds, and mammals).

    Embryonic terms for kidney types are *pronephros* for the tubule type found at the most cranial end of the holonephric kidney and the *mesonephros*, which is roughly the embryonic equivalent of the adult opisthonephros. The pronephros is formed by the evagination of the wall of the embryonic pleuroperitoneal cavity. After the pronephric tubules form, their tips fuse and initiate the *archinephric duct,* which then grows caudally through the prospective kidney tissue to the cloaca. Mesonephric and opisthonephric tubules form as evaginations of the developing archinephric duct.

    Opisthonephros is the "behind kidney" (opistho-, Greek = behind) because it lies behind the pronephros, which is the "before kidney" (pro-, Greek = before). This complication of terms is more confusing than it needs to be because it was originally thought that the pronephros, mesonephros, and metanephros succeeded one another phylogenetically as they do embryonically. When it was determined that this was not the case, it was necessary to find new terms for the adult phylogenetically successive structures.

    Pronephric tubules open by funnellike *nephrostomes* from the coelom to the tubule. The adult kidney tubule has a knot of blood capillaries, the *glomerulus,* that is encapsulated by the end of the opisthonephric tubule, the *renal* (or Bowman's) *capsule* (fig. 45).

2. *Archinephric ducts* drain the paired kidneys and open into the cloaca through the *urinary papilla*. These ducts are present in both the embryonic (pronephros and mesonephros) and adult (opisthonephros) kidneys of both sexes. In the adult male the pronephric tubules are associated exclusively with the testes, and the rostral portion of the archinephric duct is also an exclusively reproductive structure in the adult male. The major portion of the duct serves to transport both sperm and kidney excretions in the male.

    In the adult female all of the embryonic pronephros is absent, and the archinephric duct is both shorter and smaller than in the male. To find the female archinephric ducts, carefully remove the pleuroperitoneal membrane and the oviduct from the surface of the kidney. The slender and

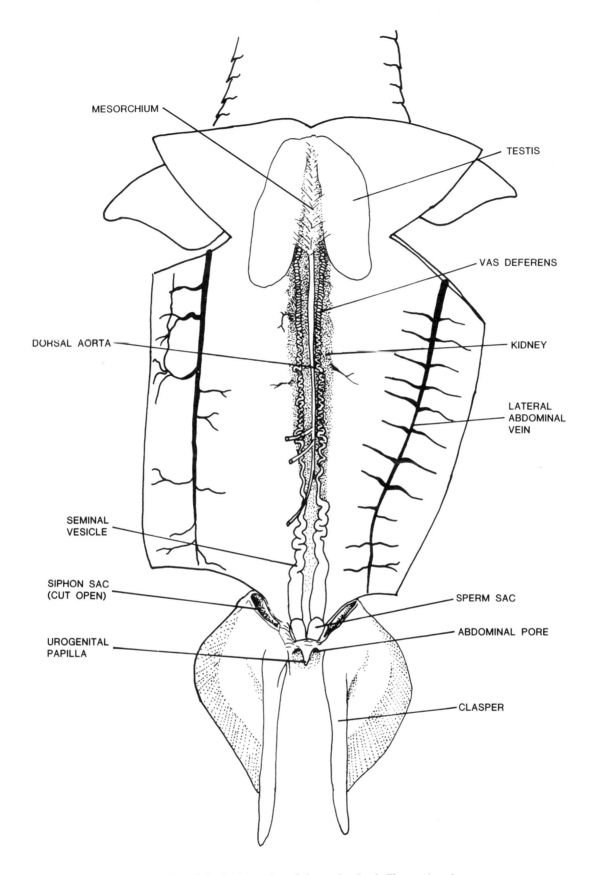

FIGURE 44. Ventral view of the body cavity of the male shark illustrating the urogenital system.

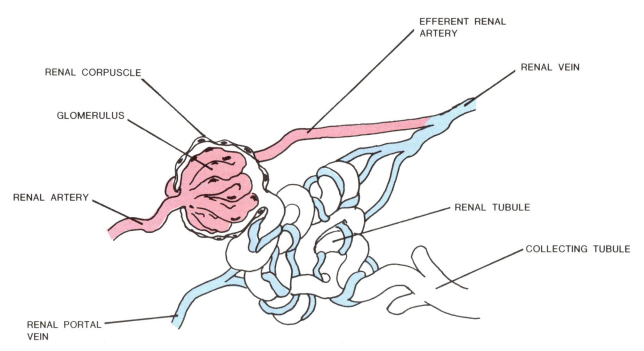

FIGURE 45. Diagrammatic reconstruction of a kidney tubule of the shark and its blood supply.

fragile archinephric duct lies on the ventral surface of the kidney dorsal to the oviduct. These ducts might be more easily traced from the urinary papillae in the cloaca and carefully working forward to the kidney by gently pulling on the papilla. Observe the taut duct and pick away the kidney tissue around the duct.

The male archinephric duct is enlarged and strongly coiled to form the *vas deferens* on the ventral surface of the kidney. Notice this duct is expanded into a *seminal vesicle* near its caudal end (fig. 44).

3. *Accessory urinary ducts* drain the more caudal portion of the kidney and are best seen in males dorsal to the seminal vesicles. At best these ducts are difficult to find and are usually absent in females.

4. The *urinary papilla* receives both the right and left archinephric ducts as well as the accessory urinary ducts, if present. The projection of the urinary papilla into the cloaca has already been noted. In males this papilla is sometimes called a "urogenital papilla" because it receives the sperm sacs of the seminal vesicles as well as the urinary ducts. The cavity of the urogenital papilla is a "urogenital sinus."

5. The *cloaca* is a common chamber receiving the openings of the urinary (or urogenital) papilla, the uteri (of the female), and the intestine (the anus). The urinary papilla opens to the dorsal half of the cloaca, the *urodeum,* which is partially separated from the ventral *proctodeum* by a low longitudinal ridge (fig. 46B).

6. The *rectal gland* is a fingerlike extension of the rectum supported by the mesorectum; served by the rectal gland artery, a branch of the caudal mesenteric artery; and drained by the rectal gland vein into the dorsal intestinal vein. The gland secretion is not under neural control although the muscles and blood vessels of the gland are enervated by postganglionic fibers of the sympathetic nervous system.

The rectal gland is a tubular gland with a central lumen. The gland cells concentrate sodium chloride (NaCl) and secrete this salt into the gland tubule. From the tubule the salt is passed into the central cavity (lumen) of the gland and from here, through a short duct, into the rectum.

SUGGESTED READINGS

Bates, G. A. 1914. The pronephric duct in elasmobranchs. *J. Morphol.* 26: 345–73.

Burger, J. W., and W. N. Hess. 1960. Function of the rectal gland in the spiny dogfish. *Science* 131:670–71.

Chan, D. K. O., and J. G. Phillips. 1967. The anatomy, histology, and histochemistry of the rectal gland in the lip-shark *Hemiscyllium plagiosum* (Bennett). *J. Anat.* 101:137–57.

Forster, R. P., L. Goldstein, and J. K. Rosen. 1972. Interrenal control of urea reabsorption by renal tubules of the marine elasmobranch, *Squalus acanthias*. *Comp. Biochem. Physiol.* 42A:3–12.

Goldstein, L., and P. J. Palatt. 1974. Trimethylamine oxide excretion rates in elasmobranchs. *Am. J. Physiol.* 227: 1268–72.

Pang, P. K. T., R. W. Griffith, and J. W. Atz. 1977. Osmoregulation in elasmobranchs. *Amer. Zool.* 17:365–77.

Schmidt-Nielsen, B., B. Truniger, and L. Rabinowitz. 1972. Sodium-linked urea transport by the renal tubule of the spiny dogfish *Squalus acanthias*. *Comp. Biochem. Physiol.* 42A:13–25.

Chapter 9
Reproductive Systems

The spiny dogfish is *viviparous* in the elementary sense that the eggs are retained and embryonic development is completed *in utero*. Nourishment for the developing embryo is derived from the yolk of the egg. There is no placenta in this species to provide nourishment for the embryo from the circulating fluid of the mother.

A full spectrum of reproductive patterns occur in the Chondrichthyes from the oviparous (egg laying) bullhead sharks to the placental viviparous hammerhead and requiem sharks. Between these extremes there are all possible varieties of viviparity. The spiny dogfish has the most primitive viviparous (also referred to as *ovoviviparity*) condition.

Sharks used for dissection are frequently immature and their reproductive organs will be underdeveloped. If this is true of your specimen locate as many structures as possible, then look at specimens being dissected by your classmates. This will also give you the opportunity to observe sharks of a different sex than your specimen.

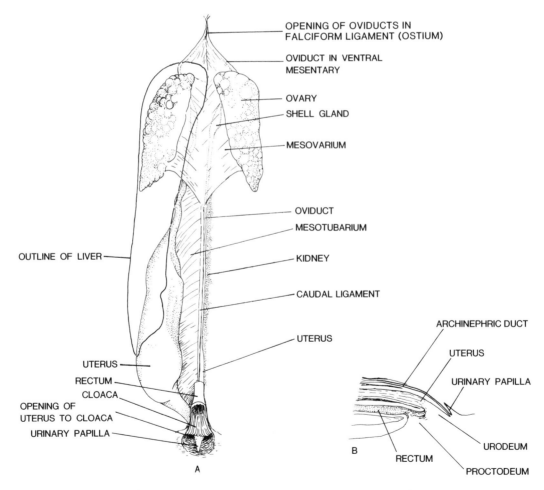

FIGURE 46. Ventral diagrammatic views of the pregnant female shark (*left*) and nonpregnant (*right*) female genital system. Inset figure *B* is a diagrammatic sagittal view of the female cloaca.

MESENTERIES OF THE REPRODUCTIVE SYSTEMS

These membranes were noted earlier (p. 30) but should be examined in more detail at this time. Locate and examine the following:

1. The *mesorchium* is the mesentery of the testis. Genital arteries, veins, and nerves traverse this membrane enroute to, and from the testis.
2. The *mesovarium* is the mesentery of the ovary. This mesentery serves to attach the ovary to the dorsal body wall as well as to support the genital arteries, veins, and nerves.
3. The *mesotubarium* is the mesentery of the oviduct, shell gland, and caudally placed uterus. This mesentery is undeveloped in immature females and the oviduct is pressed tightly against the kidney surface.

FEMALE

1. The *ovaries* are two soft, oblong, cream-colored glands, beside and dorsal to the lobes of the liver. Several somewhat hemispherical swellings may occur on the dorsal side—these are eggs in various stages of development. Immature ovaries lack visible eggs.
2. The *ostium* is a vertical slit between two layers of the *falciform* (sickle-shaped) *ligament* located in the midline at the cranial end of the body cavity. This slit, closed in immature females, opens from the pleuroperitoneal cavity into the oviducts to admit eggs discharged from the ovaries into the pleuroperitoneal cavity. If the ostium cannot be probed, make a transverse slit in the left oviduct about two inches dorsolateral to the ostium, and pass a curved probe through the tube until it emerges at the ostium.
3. The *oviduct* is the tube in which the ova (eggs) are usually fertilized (sharks have internal fertilization—most fishes do not). Follow the left oviduct dorsal to the ovary and observe a thick-walled shell gland.
4. The *shell gland* (=nidamental gland) secretes a thin membrane around several eggs at a time in sharks.
5. The *uterus* is the much enlarged caudal portion of the oviduct. It is here that fertile eggs undergo gestation for from sixteen to twenty months after which the young dogfish "pups" are born. Since the embryo is nourished by an ample yolk, there appears to be no need for a placenta. This type of development has been termed *ovoviviparous*. However, some investigators believe delicate uterine villi may contact the "shell" membrane covering the embryos and contribute some nourishment for their development.

MALE

1. The *testes* are the male gonads. They are soft, slender glands situated in positions comparable to those occupied by the ovaries. Look for the testes dorsal to the liver lobes. These reproductive glands contain many microscopic tubules, the lining (germinal epithelium) of which produces the sex cells (spermatozoa).
2. The *efferent ductules,* five or six in number, pass through the mesorchium from the testis to the cranial part of the kidney (=epididymis) and carry sperm from the testis to the epididymis. Hold up the testis to stretch the mesorchium, and looking toward the light, try to determine how many efferent ductules pass from one testis to the kidney.
3. The *epididymis* is a region of sperm conducting tubules represented by the cranial end of the kidney whose tubules have little, if any, urinary function. Efferent ductules from a testis connect with the tubules of the epididymis, and these in turn join the opisthonephric duct, which functions as a sperm duct in males. In a mature male the opisthonephric duct is greatly convoluted, but in young males it is straight like the corresponding duct in females.
4. The *vas* (or ductus) *deferens* (see archinephric ducts, chapter 8, no. 2.) is a much coiled tube on the ventral side of the epididymis and opisthonephros. That portion of the vas deferens lying immediately caudal to the testis is sometimes known as *Leydig's gland* for it is believed to secrete a fluid substance necessary to the normal function of the spermatozoa.
5. The *seminal vesicle* is an enlarged and straightened caudal section of the vas deferens through which sperm cells pass and from which secretions may be added to the contents of the tube. Dissect away the peritoneum from the ventral face of the left kidney, and trace the seminal vesicle to the cloaca.
6. The *sperm sacs* are a pair of small sacs evaginated from the caudal ends of the seminal vesicles. Find them on either side of the cloaca where they project cranially and lie parallel to the ventral sides of the seminal vesicles. Open one of these sacs and identify the small papilla formed by the terminus of the seminal vesicle.
7. The *urogenital papilla* is large and sometimes curved in mature males. Since the uteri of the female open to the cloaca separately from the archinephric duct, the papilla is a smaller and uncurved, *urinary papilla* in the female. If the cloaca has not been opened yet, slit it open with a short longitudinal cut and spread the walls

apart. Identify the *urodeum,* which receives the urogenital (or urinary) papilla, and the *coprodeum,* which receives intestinal wastes. These two chambers join to form a *proctodeum,* which opens externally. Now slit open the urogenital papilla, and observe that it contains a small cavity, the *urogenital sinus,* into which the two sperm sacs open caudally. Both urine and sperm cells enter the cloaca through the terminal pore of this papilla.

8. *Siphon sacs* are paired blind sacs located on either side of the midline just cranial to the cloaca (fig. 44). These sacs are subdermal pockets opening at their caudal end into the clasper groove. They are sheathed with a thin muscle layer and lined with an epithelium containing secretory cells that produce serotonin that causes powerful contractions in the female uterus.

The siphon sacs are filled with seawater by the flexing movements of the claspers. During copulation, the muscular wall of the sac contracts forcing the seawater through the clasper groove thus washing the sperm into the female oviduct.

Locate a siphon sac by cutting transversely through the skin at the cranial end of the pelvic fin. Insert a probe caudally through the sac until it emerges into the clasper groove.

9. *Clasper groove* (=spermatic sulcus) is on the dorsal side of the clasper where it functions as a narrow trough to convey the sperm to the caudal end of the clasper during copulation.

SUGGESTED READINGS

Gilbert, P. W., and G. W. Heath. 1972. The clasper-siphon sac mechanism in *Squalus acanthias* and *Mustelus canis. Comp. Biochem. Physiol.* 42A:97–119.

Hisaw, F. L., and A. Albert. 1947. Observations on the reproduction of the spiny dogfish, *Squalus acanthias. Biol. Bull.* 92:187–99.

Jones, N., and R. C. Jones. 1982. The structure of the male genital system of the Port Jackson shark *Heterodontus portusjacksoni* with reference to the genital tracts. *Australian Jour. Zool.* 30 (4): 523–42.

Mattisson, A., and R. Faenge. 1982. The cellular structure of the Leydig organ in the shark *Etmopterus spinax. Biol. Bull.* (Woods Hole) 162 (2): 182–94.

Schlernitzauer, D. A., and P. W. Gilbert. 1966. Placentation and associated aspects of gestation in the Bonnethead shark. *Sphyrna tiburo. J. Morph.* 120 (3): 219–32.

Stanley, H. P. 1971. Fine structure of spermiogenesis in the elasmobranch fish *Squalus suckleyi.* II. Late stages of differentiation and structure of mature spermatozoon. *J. Ultrastruct. Res.* 36: 103–18.

Tanaka, S., M. Hara, and K. Mizue. 1979. Studies on sharks, Part 13, Electron microscopic study of spermatogenesis of the squalid shark *Centrophorus atromarginatus. Jap. J. Ichthyol.* 25 (3): 173–80.

Wourms, J. P. 1977. Reproduction and development in chondrichthyan fishes. *Amer. Zool.* 17: 379–410.

Chapter 10
Nervous System

The nervous system receives external stimuli through both general and special sense organs and activates muscles or glands with impulses elicited by nerves. The stimuli are coordinated to produce the correct impulse by the central nervous system (spinal cord, of all chordates, and brain of vertebrates). Impulses are then carried by nerves of either the: (1) peripheral (somatic) or (2) autonomic (sympathetic and parasympathetic nerves) system.

The nervous system is thus a communications system closely allied to the endocrine system, both of which communicate by chemical messengers.

In this chapter we will deal with the nerves of the peripheral and central (spinal cord and brain) nervous systems.

THE CENTRAL NERVOUS SYSTEM

The Brain

The brain of vertebrates is formed embryonically from three primary lobes: *prosencephalon, mesencephalon,* and *rhombencephalon.* During development these lobes become separated into five subdivisions and each of these five subdivisions give rise to the fully developed structures seen on the adult brain. Figure 47 will indicate the relationship of the various divisions, structures, and ventricles of the shark brain to each other. A dorsal, *hollow* nerve cord is a primary chordate characteristic. That is, all members of the phylum chordata have a central cavity in their spinal cord; vertebrates in addition have cavities in their brain (ventricles) that are continuous with the central cavity of the spinal cord. The central cavities or ventricles are filled with *cerebrospinal fluid,* which also bathes the outside of the brain and spinal cord.

The brain and spinal cord are surrounded by an outermost membrane, the *dura mater,* which adheres to the inner wall of the chondrocranium and to the neural canal of the vertebrae. A thin, delicate membrane, the *pia mater,* lies on the surface of the brain. The pia mater and dura mater are connected to each other by thin delicate fibers, which, in some other vertebrates, form an *arachnoid* layer. The cavity between the pia and dura mater and containing the arachnoid fibers is filled with cerebrospinal fluid.

Dissection

If separate shark heads are available, one half of the class should perform the ventral view dissection and the other half should dissect the dorsal view. Your instructor may also decide to divide the class in this manner to dissect your original specimen. If you do perform both dorsal and ventral view dissections on the original specimen, do the ventral view first. If a separate head is used for the brain dissection the original specimen should be saved for dissection of the ear and eye.

Avoid the eye and ear as much as possible during your dissection of the brain.

Ventral View

To expose the ventral surface of the brain remove the lower jaws completely. Next, scrape the soft tissues including the efferent branchial arteries off the roof of the mouth. This will expose most of the basal plate of the chondrocranium. Use a scalpel and cut the tissues holding the upper jaw to the chondrocranium. Next, cut the orbital process on the side of the upper jaw and remove the upper jaw and the tissues on the ventral side of the rostrum. Leave the eye in place in the orbit and the nerves that are now exposed should be left intact.

Carefully shave away the cartilage of the basal plate until you reach the cranial cavity. Place one tip of a blunt forceps carefully in the cranial cavity beneath the cartilage and pull the piece of cartilage away from the cranium. Continue this procedure until you expose the ventral surface of the brain. The cranial nerves pass out of the chondrocranium through the lateral walls. Pick away these lateral walls carefully to expose the nerves. Use a probe or forceps so you do not cut the nerves. Cranial and peripheral nerves are tough but they are easily cut and they

(Text continues on page 66.)

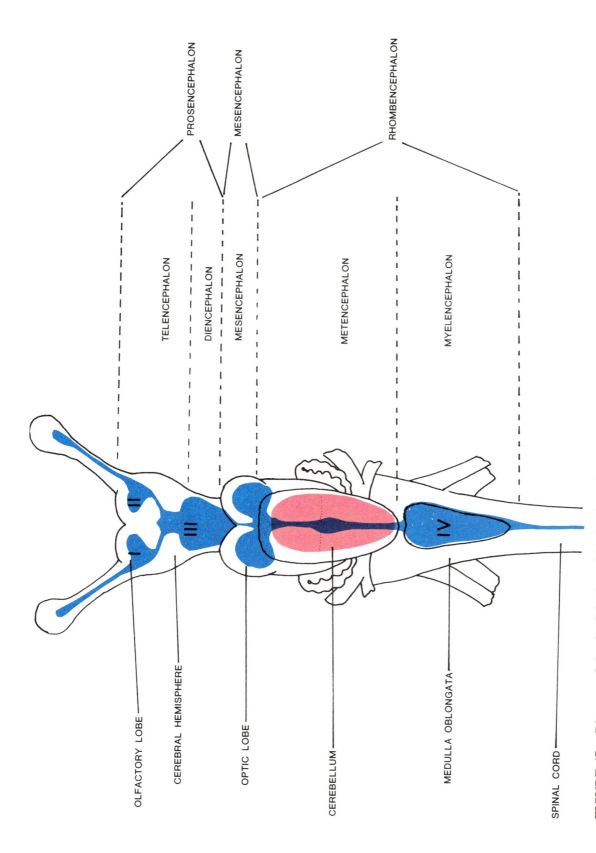

FIGURE 47. Diagram of the shark brain and its divisions. Primary and secondary subdivisions are indicated on the *right* and major structures are labeled on the *left*. Roman numerals indicate principal ventricles of the brain and colored areas include the ventricles and canals of the brain.

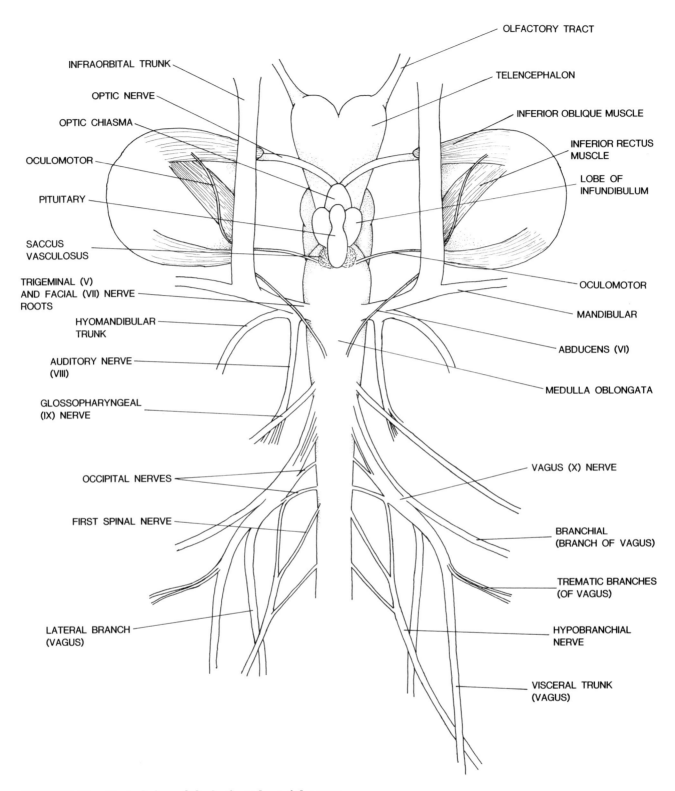

FIGURE 48. Ventral view of the brain and cranial nerves.

CHART OF THE CRANIAL NERVES

NUMBER	NAME	SUPERFICIAL ATTACHMENT ON BRAIN	FUNCTION	DISTRIBUTION
0	Nervus Terminalis	Mid rostral border of olfactory lobes	Sensory ?	Olfactory bulb.
I	Olfactory	Lateral rostral border of olfactory lobes	Sensory	From epithelium of olfactory sac.
II	Optic	Midventral surface of diencephalon rostral to pituitary	Sensory	From retina of eye.
III	Oculomotor	Ventrolateral surface of mesencephalon near caudal end of pituitary	Motor	To superior, medial, and inferior rectus muscles and inferior oblique muscle of the eye. A small branch of this nerve also serves the protractor lentis muscle that moves the lens (see p. 78).
IV	Trochlear	Dorsolateral surface of mesencephalon between the optic lobes and cerebellum	Motor	To superior oblique muscle of the eye.
V	Trigeminal	Ventrolateral surface of the rostral medulla oblongata with facial and auditory	Motor and Sensory	Four major branches: 1. Superficial ophthalmic (with Facial, VII)—sensory from skin of the rostrum. 2. Deep ophthalmic—sensory from skin of the rostrum. 3. Infra-orbital trunk (with Facial, VII)—sensory from skin of the rostrum. 4. Mandibular—sensory from skin and motor to muscles of the jaws.
VI	Abducens	Ventromedial surface of medulla just caudal to trigeminal	Motor	Lateral rectus muscle.
VII	Facial	Ventrolateral surface of rostral medulla with trigeminal and auditory	Motor and Sensory	Four major branches: 1. Superficial ophthalmic—sensory from dorsorostral lateral line and Lorenzini canals. 2. Buccal—sensory from ventrorostral lateral line and Lorenzini canals. 3. External mandibular (Hyomandibular)—sensory from lateral line and Lorenzini canals around mouth and from oral cavity. 4. Palatine—sensory from taste buds of the oral cavity.

VIII	Auditory	Caudal surface of trigeminal-facial trunk on the ventrolateral surface of medulla	Sensory	Maculae and cristae of the membranous labyrinth of the ear.
IX	Glossopharyngeal	Ventrolateral surface of medulla just caudal to auditory nerve	Motor and Sensory	Sensory from and motor to pharyngeal musculature regulating respiratory movements of the pharynx. Sensory from rostral portion of caudal lateral line canal.
X	Vagus	Ventrolateral surface of the caudal ⅓ of medulla oblongata by numerous roots	Motor and Sensory	Three major divisions: 1. Lateral branch—sensory from lateral line canal (smallest division of the main vagal trunk). 2. Visceral branch—sensory and motor from and to the four posterior gill pouches; motor to the cucullaris muscle; sensory and motor from and to the heart and digestive tract. (This is the largest, and most complex division of the vagus trunk). 3. Hypobranchial (=accessory nerve of amniotes)—sensory from skin and motor to hypobranchial musculature. Has some fibers with vagus trunk but mainly formed by two occipital nerves from caudal medulla and first three spinal nerves from rostral spinal cord. This nerve may also be considered the XI cranial nerve or spinal accessory nerve. The XII, Hypoglossal nerve of amniotes is represented in sharks by fiber tracts in parts of the visceral trunk and hypobranchial nerves.

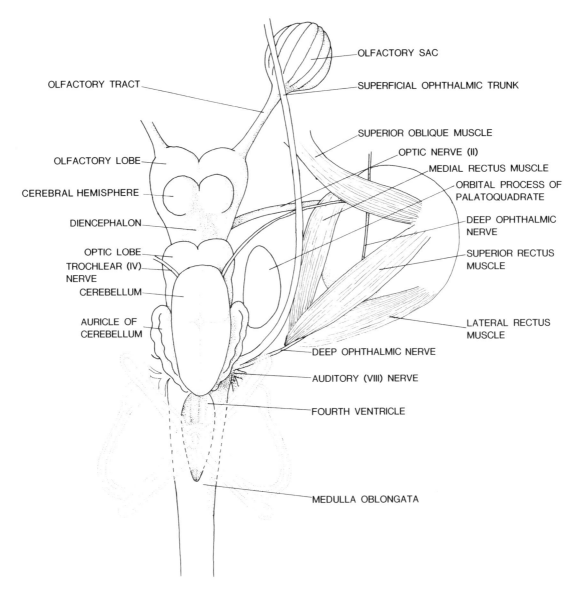

FIGURE 49. Dorsal view of the brain.

pull loose easily from their attachment on the brain. Consequently, use great care to avoid cutting or pulling the nerve *away* from the brain. On the other hand you can pull the nerve toward the brain with relative impunity.

Review the internal carotid circulation to the brain (fig. 39) when the ventral aspect is exposed.

1. The *telencephalon* in ventral view is not easily divisible into olfactory and cerebral components. These hemispheres will best be seen in the dorsal view of the brain (fig. 49).
2. The *optic chiasma* is the crossing of the optic nerve fibers from the retina of the eye to the opposite side of the brain (figs. 48 and 50), that is, all the fibers from the right eye pass to the left side of the brain and those from the left eye pass to the right side of the brain. After crossing, the optic nerve fibers become the optic tract in the brain and pass to the optic lobes (which are best seen in a dorsal or lateral view).
3. The *pituitary gland* is a complex structure formed from two different embryonic areas. The *infundibulum* and *pars nervosa* are evaginations from the ventral diencephalon, and the *adenohypophysis,* consisting of a *pars distalis* and a *pars intermedia,* evaginates from the roof of the embryonic mouth. The shark pituitary is further complicated by a bilobed infundibulum that terminates in an expanded vascular sac (figs. 48, 50, and 51). Between the two lobes of the infundibulum, the gray matter in the floor of the third ventricle is the *hypothalamus,* and the fiber tracts on the infundibulum ending at the *saccus vasculorum* represents the *pars nervosa.* Hormones

are released by the pars nervosa to the general circulation via the saccus vasculorum. Other hormones are released by cells of the hypothalamus to a pituitary blood portal system that delivers the secretions to the adenohypophysis.
4. The *medulla oblongata* is the tapering caudal portion of the brain extending from the caudal end of the pituitary gland to the spinal cord. In ventral view, the medulla is a straight tapering structure with the majority of the cranial nerves entering or leaving its surface (fig. 48).
5. *Cranial nerves*—All except the *Trochlear* (IV) cranial nerve may be seen in a ventral view. Refer to the Chart of the Cranial Nerves (pp. 64 and 65) for a description of these nerves.

Dorsal View

Expose the dorsal surface of the brain by shaving thin layers off the chondrocranium covering it, but be careful not to cut any nerves passing through the cartilage.

This dissection will also expose the eye and labyrinth of the ear. Be especially careful that you do not destroy the delicate membranous semicircular canals of the ear and leave sufficient cartilage in place around the canals to serve as support.

1. *Olfactory bulbs* are paired extensions of the brain that lie in contact with the olfactory sacs (fig. 49) that contain the olfactory sensory epithelium.
2. The *olfactory tract* is a stalk or peduncle of olfactory nerve fibers between the olfactory bulb and the rostral end of the brain (figs. 49 and 51).
3. The *olfactory lobes* are the rostral halves of the telencephalon. They are separated from the smaller *cerebral hemispheres* (or lobes) by a slight transverse groove (figs. 49, 50, and 51).
4. The *cerebral hemispheres* are a pair of dorsal "swellings" of the telencephalon each containing a cavity, the first or second ventricle of the brain (see fig. 47).
5. The *diencephalon* (between brain) lies between the fore- and mid-brains. It appears superficially as a connection between the cerebrum and optic lobes, but internally, the walls have important gray matter forming the thalamus and hypothalamus. The roof is a highly vascularized membrane, the *choroid plexus* (fig. 51).
6. The *optic lobes* are the most prominent part of the mesencephalon. These lobes actually represent the tectum as well as the optic lobe of higher vertebrates (figs. 50 and 51).
7. The *cerebellum* is a large mass lying caudal to the optic lobes. It has faint transverse and longitudinal grooves that divide it into four parts (figs. 49, 50, and 51).
8. The *auricles* (=little ears) or *restiform bodies* of the cerebellum are a pair of "earlike" nerve tracts at the junction between the cerebellum and the *medulla oblongata* (figs. 49 and 50).
9. The *medulla oblongata* is the elongated caudal part of the brain tapering caudally to the spinal cord. Most of the roof of the medulla oblongata is formed of *tela choroidea* with projection into the fourth ventricle (the cavity of the medulla oblongata). The tela choroidea, together with blood capillaries, forms a choroid plexus similar to that in the roof of the other ventricles.
10. The *fourth ventricle* can be seen after removing the tela choroidea and auricles (figs. 50 and 51). Lift the caudal end of the cerebellum and note the nerve fiber tract connecting the auricles with each other and with the cerebellum proper. The inner wall of the fourth ventricle has longitudinal nerve columns that may be identified:
    a. *Somatic sensory column* is the most dorsal of the four columns, forming a strip on the dorsal lip of the ventricle. The rostral part of this tract is continuous with the auricles and is called the *acoustico-lateralis* area.
    b. The *visceral sensory column* is just ventral to the somatic sensory column and conveys sensations from the viscera to the brain.
    c. The *visceral motor column* parallels the visceral sensory column as a slender tract just ventral to the latter.
    d. The *somatic motor column* is a prominent nerve tract lying in the floor of the fourth ventricle. These paired tracts carry impulses for the contraction of the skeletal (voluntary) muscles while the visceral motor fibers innervate the smooth (involuntary) muscles of the viscera.

Lateral View

After examining the brain and cranial nerves in ventral and/or dorsal views, cut the cranial nerves, leaving a stub of each nerve attached to the brain. Next, transect the spinal cord near the medulla oblongata and carefully remove the brain from what remains of the cranium. Reexamine the brain in ventral and dorsal views and now in lateral view as follows:

1. *Olfactory lobe* in lateral view is the ventrorostral portion of the telencephalon (fig. 50).
2. The *cerebral hemispheres* (together the *cerebrum*) are the dorsal most rounded knobs above the olfactory lobes (fig. 50).
3. The *diencephalon* is the trunk area between the cerebrum and optic lobes. Ventrally, the major feature of this region is the optic chiasma and optic nerves.

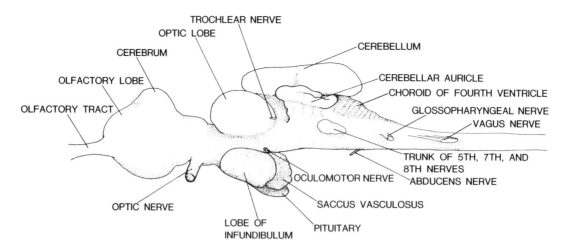

FIGURE 50. Lateral view of the brain.

4. The *epiphysis* or *pineal* apparatus will have been lost during your dissection of the brain. It is doubtful that an adequate dissection demonstrating the pineal can be performed on the shark. Reconstruction of this organ may be accomplished from histological studies of this region by cutting out a block of the area containing the pineal from the intact shark head and preparing sections of the region for microscopic examination.
5. The *optic lobes* are the rounded dorsal portions of the mesencephalon (figs. 50 and 51).
6. The *pituitary* complex is a projection of the ventral diencephalon lying directly ventral to the optic lobes (figs. 50 and 51).
7. The *cerebellum* is the large "mushroom-shaped" structure just caudal to the optic lobe. The cerebellar auricle may be seen in this view. The cerebellum and its auricles form the dorsorostral portion of the rhombencephalon (figs. 50 and 51).
8. The *medulla oblongata* is the ventrocaudal portion of the rhombencephalon.

Sagittal Section

Place the shark brain ventral side down on a surface that will not be damaged by cutting with a razor blade or scalpel (DO NOT place on formica or similar surface). Use as long a blade as you can find and cut the brain in halves by drawing the blade in a single stroke from anterior, between the olfactory lobes, to the posterior tip, passing through the middle of the cerebellum and medulla. If the blade is too short to completely divide the brain, carefully spread the cut upper portions and make a second cut similar to the first to divide the brain. Keep the two sections moist in a dish of water and store moist between laboratory sessions.

1. The cavity of the telencephalon represents one of the two (first and second) *ventricles* of the forebrain. Cut the cerebral hemisphere and trace one of these ventricles along the olfactory tract to the olfactory bulb.
2. The *foramen of Munro* (interventricular foramen) is the passage between the lateral (first or second) ventricle and the *third ventricle*, which is contained in the diencephalon and extends into all parts of the hypothalamus.
3. The *hypothalamus* forms the floor of the diencephalon and includes the following:
   a. The *infundibulum* with its paired inferior lobes and saccus vasculosus.
   b. The *mammillary bodies* form the thickened portion of the diencephalic floor just above the saccus vasculosus.
4. The *thalamus* is the gray matter of the lateral wall of the diencephalon and is one of a bilateral pair of important correlation centers in the brain.

   The roof of the diencephalon contains the *habenula*, the *pineal body*, a *choroid plexus*, and a *velum transversum*.
   a. The habenula is a small brain center appearing as a thickening between the optic lobes and the roof of the diencephalon.
   b. The pineal body extends dorsorostrally from the habenula and is usually lost during dissection.
   c. The *tela choroidea* is a thin roof over the diencephalon that folds down into the third ventricle to form the choroid plexus with a blood vascular bed. The rostral part of this tissue forms a sac called the *paraphysis* in the shark.
   d. Caudal to the paraphysis is a thin transverse partition, the *velum transversum,* marking the boundary between the telencephalon and the diencephalon.

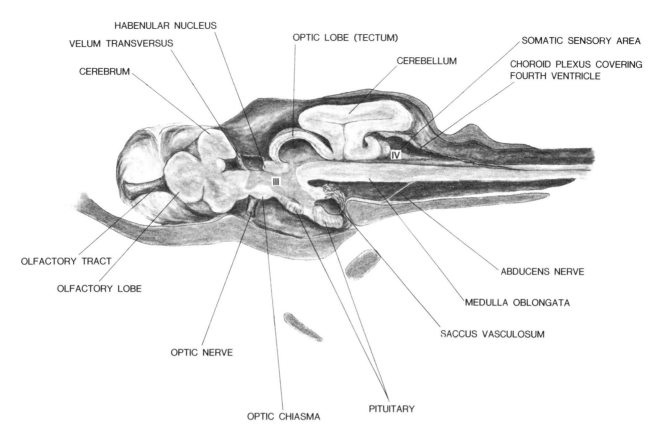

FIGURE 51. Sagittal section of the shark brain.

5. Passing caudal from the third to the fourth ventricle is a narrow passage called the *cerebral aqueduct,* or *aqueduct of Sylvius,* that provides for drainage of cerebrospinal fluid from the third to the fourth ventricle. The aqueduct also communicates with the optic and cerebellar ventricles.
6. The *fourth ventricle* is the cavity of the medulla oblongata. The floor of the ventricle is known as the *fossa rhomboidea,* and the roof is a choroid plexus composed of a tela choroidea and a vascular network.
7. *Cerebrospinal fluid* is secreted by the choroid plexi into the ventricles of the brain. This lymph-like fluid helps nourish both the internal and external surfaces of the brain and to maintain a constant brain temperature. The fluid moves from the lateral ventricles through the *foramen of Munro* to the third ventricle and from here through the cerebral aqueduct to the fourth ventricle and central canal. The fluid exits the fourth ventricle through minute openings in the choroid plexus and passes into the subdural spaces covering the brain and spinal cord. The fluid is absorbed back into the blood through blood sinuses in the dura mater.

SPINAL CORD

Spinal nerves arise from the spinal cord by two roots at regular intervals along the length of the spinal cord. The dorsal root passes through a foramen in the dorsal intercalary plate anterior to the ventral root, and the ventral root passes through the neural plate foramen. The two roots fuse outside the neural canal (fig. 52) just beyond the dorsal root ganglion, which is also outside the neural canal. Branches of the spinal nerves serve the muscles of the trunk and fins.

All of the sensory nerve cell bodies with their nuclei are located in the ganglion of the dorsal nerve root. The nerve cell bodies of the motor neurons are located in the gray matter of the spinal cord. In those fishes that have been studied, some of the efferent visceral motor fibers pass through the dorsal root. All other efferent motor fibers pass out of the spinal cord through the ventral spinal nerve root.

Within the spinal cord the gray matter containing nerve cell bodies and synaptic junctions, is located in the center of the cord. When cut in cross section the gray matter is roughly shaped like the letter H with dorsal and ventral "horns." The central canal of the spinal cord is in the center of the gray matter.

The white matter consists of myelinated nerve fibers arranged in tracts. The tracts between the two dorsal horns of gray matter are called the *dorsal funiculi.* The fibers

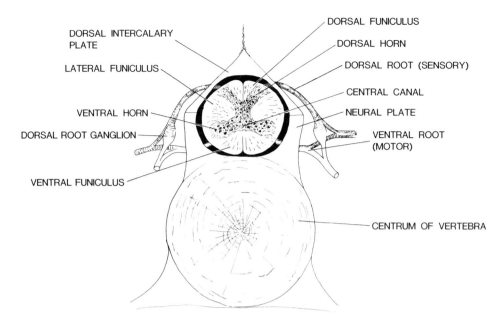

FIGURE 52. Cross section of the spinal cord and vertebra to illustrate the arrangement of the spinal nerves and gray and white matter of the spinal cord.

are primarily sensory and carry impulses toward the brain. The *ventral funiculi* are between the ventral horns of gray matter and are primarily motor carrying impulses caudally, away from the brain.

Spinal nerves serving the fins fuse into a plexus before branching into the muscles proper. As pointed out earlier the first two spinal nerves join with branches of the vagus to form the hypobranchial (see Chart of the Cranial Nerves). The third through the eighth spinal nerves and branches of the hypobranchial form the cervicobrachial plexus with branches to the levator and adductor muscles of the pectoral fins. In addition, the ninth to fourteenth spinal nerves send nerves to the pectoral fin but they do not join the cervicobrachial plexus.

The cervicobrachial nerve plexus may be seen on the dorsal body wall after removal of the cardinal sinus or laterally by separating the fin muscles from the trunk muscles at the ventral base of the fin.

The pelvic fins are served by branches of the lumbrosacral plexus, which is formed by branches of the last six to ten spinal nerves of the trunk. The first (most anterior) of the spinal nerves entering the pelvic fin has nerve fibers from several more anterior spinal nerves and is therefore termed the *collector nerve*. The most caudal trunk spinal nerves are joined in a true lumbrosacral plexus before the branches enter the fin.

To find this plexus, peel the skin away from the dorsal fin base and carefully separate the trunk muscles from the pelvic levators with a blunt probe. Note that the nerves merge together at the base of the fin and fan out again as the enter the fin.

SUGGESTED READINGS

Kappers, C. U. A., G. C. Huber, and E. C. Crosby. 1936 (reprinted 1965). *The comparative anatomy of the nervous system of vertebrates, including man.* New York: The Macmillan Company. (Reprint by Hafner Publishing Company, New York, in 3 vols.)

Norris, H. W., and S. P. Hughes. 1920. The cranial, occipital and anterior spinal nerves of the dogfish. *J. Comp. Neurol.* 31: 293–395.

Chapter 11
Special Sense Endings

Sense endings are receptors that "test" the animal's environment. Many sense endings are too small to be seen by gross examination. These small sense endings may receive noxious stimuli that the animal could avoid or attractive stimuli. Most receptors are encapsulated by cells that limit or specialize the stimuli to be received. Endings for touch, pressure, heat, cold, and stretch have not been described in the shark, but these endings in mammals are too small to be seen in gross dissections and are referred to as general sense endings.

Endings for smell, sight, hearing, static and dynamic equilibrium, taste, and spatial perception are special sense endings. The actual nerve endings of the special senses are also microscopic in size but (with the exception of taste) the structures that transmit or amplify the impulse may be examined grossly. Consequently, special senses to be studied here are *olfactory* (smell), *optical* (sight), *vestibular* (equilibrium or balance; the shark's ability to hear has not been established), and *lateralis* (the perception of movement of the surrounding water). Additionally the *ampullae of Lorenzini*, a suspected electrical sense ending, will also be described.

LATERAL LINE ORGAN

The name "lateral line" comes from the fact that the major caudal portion of this organ lies just superficial to the outer edge of the transverse septum that separates the epaxial and hypaxial muscles of the trunk. The series of entry pores to the canal thus form a line on the lateral surface of the trunk corresponding to the canal next to the septum.

In the head (figs. 7 and 53) the lateral line canal is more complex. The canal passes rostrally above the eye to the lateral tip of the rostrum and then turns caudally and ventrally beneath the eye. The portion of the canal rostral to the eye is termed the *supraorbital canal*. The latter actually branches to form the supraorbital canal and a ventral branch, the *infraorbital canal,* which passes between the eye and the spiracle and joins the supraorbital canal again beneath the eye. The *hyomandibular canal* passes caudally from the reunion of the infraorbital and supraorbital canals. A *buccal canal* branches forward and ventral to the supraorbital canal.

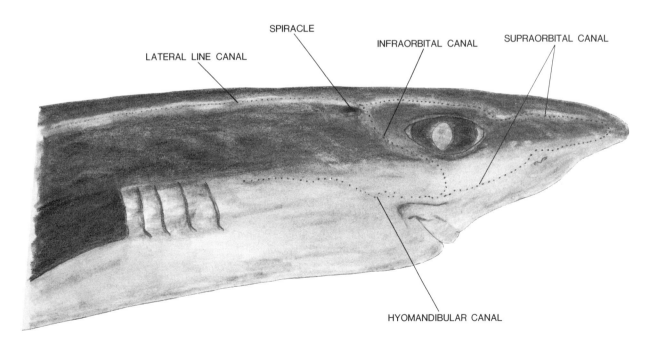

FIGURE 53. Lateral view of the shark head illustrating external characteristics.

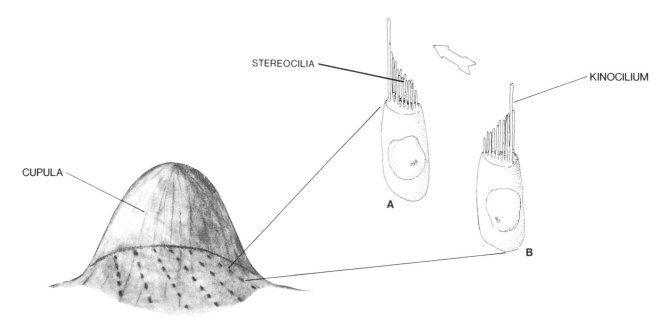

FIGURE 54. A neuromast and hair cells (*A* and *B*) of the lateral line canal system. Water movement in the direction of the arrow moves the kinocilium of *A* away from the stereocilia thus inhibiting the neural activity of the cell. By the same movement the kinocilium of *B* is moved toward the stereocilia thus stimulating cell *B*. Stimulation of one cell and inhibition of the other indicates the direction of movement to the shark.

Although you will probably not be able to observe the sensory portion of the lateral line canals, a brief description of these structures will help you to understand the workings of not only the lateral line organ but also the *cristae* and *maculae* of the ear. The sensory cells are ciliated cells arranged in a nearly continuous row on the dorsal wall of the lateral line canal. The dorsal ridge of sensory cells is called a *neuromast* (fig. 54) and is located opposite a tubule opening. Each neuromast consists of ciliated cells clumped together with their cilia embedded in a *cupula*.

Each ciliated "hair" cell has a single large *kinocilium* and numerous shorter *stereocilia*. When the kinocilium is moved away from the stereocilia the cell is excited. Movement of the kinocilium toward the stereocilia inhibits neural activity. In each neuromast the hair cells are paired so the kinocilium of one hair cell is at one side of the cell and kinocilium of its paired cell is at the opposite side. Since the cilia of both cells are embedded in the same cupula, movement of the cupula excites one cell and inhibits the other. This arrangement also occurs in the cristae of the ear ampullae and serves to determine the direction of movement.

THE EAR

The receptor cells of the ear bear a striking resemblance to those of the lateral line organ, but the structures directing the stimulus to these cells are much more complicated than the canal of the lateral line organ. The structures of the ear corresponding to the lateral line neuromasts are *cristae* in the ampullae of the semicircular canals and *maculae* in the saccule and utricle. Although you will not be able to see the cristae or maculae with a gross dissection you should find the following structures housing the neurosensory cells (figs. 55 and 56).

1. The *endolymphatic ducts* provide a common origin for the membranous labyrinths of the two (right and left) ears. The ducts begin at a common opening beneath the skin and in a depression in the dorsal caudal chondrocranium midway between the two ears. One duct passes to the right and the other to the left confluence of the three semicircular canals at the top of the *utriculus*.
2. *Perilymphatic ducts* surround the endolymphatic ducts and the external opening to the perilymphatics also surround the common opening

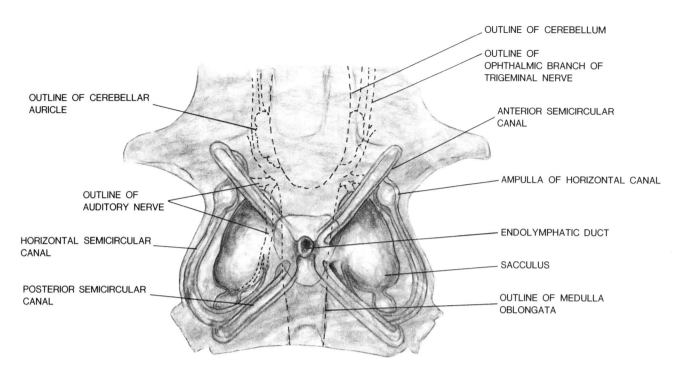

FIGURE 55. Dorsal view of the ears of the shark through a transparent chondrocranium.

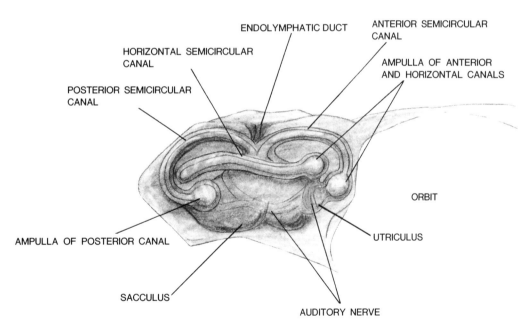

FIGURE 56. Lateral view of the ear of the shark through a transparent chondrocranium.

of the endolymphatic opening. In addition to surrounding the semicircular canals, the perilymphatic ducts open to the *vestibule,* a chamber surrounding the *utriculus* and *sacculus.* Both the endolymphatic and perilymphatic fluids are aqueous filtrates.

3. *Semicircular canals* are located in the caudolateral portion of the chondrocranium. If you have not already done so, the entire roof of the chon-

drocranium should be exposed by peeling the skin from the top of the head and cutting away the epaxial muscle attached to the dorsocaudal surface of the chondrocranium. Carefully shave away pieces of the dorsal chondrocranium with a sharp scalpel until you expose a membranous semicircular canal. Use figure 55 as a guide and determine whether this is an anterior or posterior vertical semicircular canal. Continue to cut away the chondrocranium carefully until most of the membranous vertical canals and the horizontal canal are exposed. Examine the following membranous canals:

a. *Anterior vertical semicircular canals* form a vertical arch approximately 45 degrees from the midsagittal plane. The canals are expanded into an ampulla just before opening to the ventrocaudal tip of the utriculus. The *cristae* containing the sensory "hair cells" is suspended from the dorsal wall of the ampulla.

b. *Posterior vertical semicircular canal* is in a vertical arc pointing caudally at 45 degrees from the median plane. The ampulla of this canal is at the caudal border of the utriculus just dorsal to the *lagena*. The cristae is located in the dorsal wall of the ampulla.

c. *Horizontal semicircular canal* actually has a short vertical component just internal to the arc of the posterior vertical semicircular canal. From its origin near the two vertical canals the horizontal canal arches caudally and then forms an arc in the horizontal plane ending in an ampulla containing a cristae at the rostral edge of the utriculus just opposite the ampulla of the anterior vertical canal.

4. *Vestibule* is a large chamber of the chondrocranium containing the membranous utriculus, sacculus, and lagena. The vestibule and the membranous sacs are separated from each other by perilymph and the membranous sacs contain endolymph.

5. *Membranous labyrinth* includes the membranous semicircular canals and ampullae described above, as well as the membranous sacs, utriculus, sacculus, and lagena.

a. *Utriculus* is the most dorsal of the membranous sacs and receives both openings of each of the three semicircular canals. The dorsolateral wall of the utriculus has a macula to receive stimuli when the shark is upside down so the weight of the otoliths rest on the macular cupula. The maculae are flatter and broader than the cristae and the hair cells differ in appearance. The macula hair cells of the utriculus also differ from those of the sacculus and lagena.

b. *Sacculus* is the ventrocaudal membranous sac that is broadly open to the utriculus. The macula is situated on the lateral ventral floor of the sacculus thus receiving stimuli from otoliths when the head is in a normal swimming position.

c. *Lagena* is an indistinct bulge at the caudalmost end of the sacculus. The macula of the lagena is located on the dorsocaudal wall. This chamber is much larger in reptiles and birds and is thought to be the homologue of the cochlea of mammals.

OLFACTORY ORGAN

Make a transverse cut through the rostrum (fig. 57) of the shark so your blade passes through the center of the two nostrils. This cut will section the olfactory sacs ventrally and the organs of Lorenzini dorsally, just beneath the skin.

The lining (epithelium) of the olfactory sac is in folds projecting into the cavity of the sac. The sensory olfactory cells are dispersed in the epithelium and their processes extend into the olfactory bulb where they synapse with the cell bodies of neurons that pass through the olfactory tract to the olfactory lobes of the brain.

ORGANS OF LORENZINI

These tiny organs consist of short, mucus-filled canals with the base expanded into ampullae (fig. 57). Nerve fibers contact each of the ampullae. These structures should be examined with a dissecting microscope and mounted needles.

The *organs of Lorenzini* are sensitive to low grade electrical stimuli.

THE EYE

The sensory cells of the eye are located in the retina and the function of all the other parts is to direct and concentrate light waves on the sensory cells.

Movement or adjustment of the lens from near to far vision or vice versa is called *accommodation*. Constriction or dilation of the iris regulates the amount of light entering the eye, and contractions of the ocular muscles direct the eye to receive images from various directions. The following structures are involved in these functions.

1. *Ocular muscles* are described in the Chart of the Ocular Muscles and illustrated in figure 58.
2. *Pedicel* is a cartilage peg extending from the midcaudal orbit (see fig. 13) to the external posterior point (see footnote p. 76) of the eyeball. This structure helps support the eyeball in the orbit and serves as a pivot upon which the eyeball moves when the ocular muscles contract. The orbit also contains gelatinous connective tissue to help cushion the eyeball.

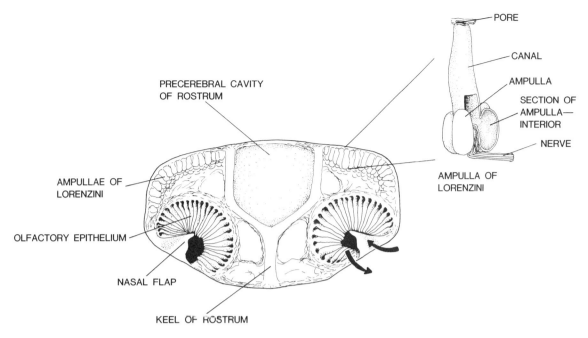

FIGURE 57. Transverse section through the olfactory sacs and rostrum of the shark showing the interior of the olfactory sac and the ampulla of Lorenzini. Arrows indicate incurrent and excurrent passageways.

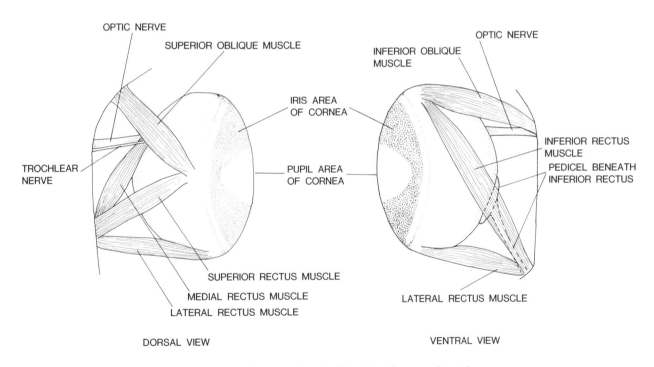

FIGURE 58. Dorsal and ventral views of the shark eyeball and ocular muscles. Also see figures 7 and 59.

CHART OF THE OCULAR MUSCLES

	ORIGIN	INSERTION	ACTION	INNERVATION*
Median (=Internal =Anterior) rectus	Ventrocaudal corner of the orbit rostral to pedicel base	Rostral margin of the eyeball	Rotates eyeball medially (rostrally)	Oculomotor (III) nerve
Lateral (=External =Posterior) rectus	Ventrocaudal corner of the orbit caudal to pedicel	Caudal margin of the eyeball	Rotates eyeball laterally (caudally)	Abducens (VI) nerve
Superior rectus	Ventrocaudal corner of the orbit dorsal to pedicel	By a thin tendon to the middorsal border of the eyeball just caudal (not posterior**) to superior oblique	Rotates eyeball dorsally and caudally	Oculomotor (III) nerve
Inferior rectus	Ventrocaudal corner of the orbit ventral to pedicel	Ventrorostral "corner" of the eyeball caudal to insertion of inferior oblique	Rotates eyeball ventrally and caudally	Oculomotor (III) nerve
Superior oblique	Rostromedial corner of the orbit lateral and ventral to the most rostral edge	By a thin tendon to the middorsal border of the eyeball just rostral to superior rectus	Rotates eyeball dorsally and rostrally	Trochlear (IV) nerve
Inferior oblique	Rostromedial corner of the orbit ventral to origin of superior oblique	Ventrorostral "corner" of the eyeball just rostral to insertion of inferior rectus	Rotates eyeball ventrally and caudally	Oculomotor (III) nerve

*See Chart of the Cranial Nerves, pp. 64–65 for details.
**Anterior and posterior in reference to the eyeball is from a point on the cornea opposite the pupil (= anterior) to the area of attachment of the pedicel to the eyeball (=posterior). Rostral and caudal refer to the long axis of the body and head (see fig. 2, pp. ix–xi).

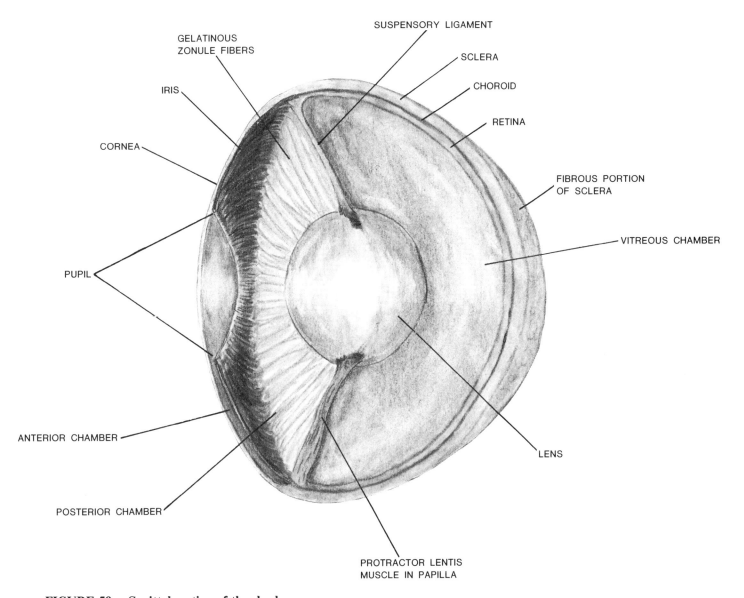

FIGURE 59. Sagittal section of the shark eye.

3. *Conjunctiva* is the transparent continuation of the epidermis over the outer surface of the eyeball. Since the conjunctiva is transparent you will not actually see this membrane. Examine the front of the eye. You are looking through both the conjunctiva and the transparent cornea just beneath the conjunctiva. The elliptically shaped area in the center of the eye is the *pupil*, an opening in the darkly pigmented *iris* that surrounds the pupil. Deep to the pupil is the *lens*. Prod the area of the pupil lightly with a dull pencil or blunt probe. The membrane you are poking is the combined conjunctiva and cornea. This also covers the iris lateral to the pupil. At the extreme rostral and caudal ends of the eye the area is white. The white area is a continuation of the cornea called the *sclera*. The sclera is an opaque tissue forming the outer covering of the eyeball.

After examining the eyeball remove the eye from the orbit by transecting each of the ocular muscles at their midpoint. Next cut the optic nerve and the pedicel. Now cut the skin around the eye and remove the eyeball.

Use a sharp scalpel and cut the eyeball into two halves by slicing from the pupil area to the attachment of the pedicel so the blade passes through the area between the insertions of the superior rectus and superior oblique dorsally and the inferior rectus and inferior oblique ventrally.

Locate the following internal structures on one of the halves using figure 59 as a guide.

4. *Cornea* and *sclera* are a single continuous outer coat of the eyeball as described above. The cornea is the clear, transparent portion covering the iris and pupil and the sclera is the tough, white fibrous portion forming the remainder of the outer coat of the eyeball.

5. *Choroid* is the darkly pigmented layer of the eye just beneath the sclera. This is continuous with the iris that lies beneath the cornea.
6. *Iris* is the thin pigmented membrane projecting toward the anterior midpoint as a continuation of the choroid. The gap in the center of the iris is the *pupil.* Sparse musculo-epithelial cells in the iris regulate the size of the pupil, but there is actually very little variation in pupil size since there is little variation in the light intensity in a shark's habitat.
7. *Anterior chamber* is the area between the iris and the cornea. In the shark this is an extremely narrow chamber and may not serve in the circulation of *aqueous humour* (the fluid of the anterior and posterior chambers).
8. *Posterior chamber* is the aqueous humour filled cavity between the iris and lens. In mammals the aqueous humour is produced by cells near the base of the iris in the posterior chamber. Without evidence to the contrary it may be assumed that this area also produces the chamber's fluid in the shark.
9. *Ciliary body* is the slightly expanded edge of the choroid layer at its border with the iris. The ciliary body proper of the shark has no muscle in it but a narrow *suspensory ligament* extends down from the middorsal point of the ciliary body to the lens and a *muscle papilla* projects up from the midventral point of the ciliary body to the lens. On either side of these midline structures the gelatinous *zonule* anchors the lens to the ciliary body.

 A small *protractor lentis* muscle from the ventral muscle papilla to the lens pulls the lens forward when the muscle contracts and allows the lens to return to its caudal position when the muscle relaxes. By moving the lens forward the eye is accommodated for near objects. In the resting position the eye is moved farther from the cornea (closer to the retina) and is focused on distant objects.

10. The *lens* is the central, spherical structure suspended to the ciliary body by a dorsal suspensory ligament and a gelatinous zonule. The shape of the spherical lens cannot be altered for accommodation as it is in reptiles, birds, and mammals.
11. *Retina* is the innermost layer of the eyeball. The retina ends at the posterior border of the ciliary body anteriorly and penetrates the sclera and choroid layers on the posterior wall of the eyeball via the optic nerve. The retina contains the sensory cells of the eye and the layer of the retina bordering the *vitreous humour* contains the nerve fibers (axons) transmitting the impulse to the brain through the optic nerve.
12. The *vitreous humour* fills the vitreous chamber between the lens and the retina. This material is more viscous than the aqueous humour of the anterior and posterior chambers and is formed into a relatively distinct body.

SUGGESTED READINGS

Bennett, M. V. L., and W. T. Clusin. 1978. Physiology of the ampulla of Lorenzini, the electroreceptor of elasmobranchs. In *Sensory biology of sharks, skates, and rays.* Arlington, Va.: Office of Naval Research, Department of the Navy, pp. 483–506.

Flock, A. 1965. Transducing mechanisms in the lateral line canal receptors. *Cold Springs Harbor Symposia on Quantitative Biology* 30: 133–45.

Gilbert, Perry W. 1962. The behavior of sharks. *Sci. Am.* 207 (1): 60–68.

Roberts, B. L. 1978. Mechanoreception and the behavior of elasmobranch fishes with special references to the acoustico-lateralis system. In *Sensory biology of sharks, skates, and rays.* Arlington, Va.: Office of Naval Research, Department of the Navy, pp. 331–90.

Walls, G. L. 1942 (reprint 1963). *The vertebrate eye and its adaptive radiation.* Bloomfield Hills, Mich.: The Cranbrook Institute of Science. (Reprinted 1963, Hafner Publishing Company, New York.)

Wersall, J., A. Flock, and Per-G. Lundquist. 1965. Structural basis for directional sensitivity in cochlear and vestibular sensory receptors. *Cold Springs Harbor Symposia on Quantitative Biology* 30:115–32.

Index

abdominal pore, 55
abducens, 63, 64, 68, 69, 76
adductor branchiae, 23, 24
adductor mandibulae, 18, 20–23
adductor muscles, 70
adductor process, 11
adenohypophysis, 66, 67
afferent branchial arch, 45
afferent branchial artery, 43–45
agnatha, viii, 36
agnathans, 54
ameloblasts, 33
amphibia, viii
amphibians, 16, 30, 40
ampulla, 73–75, 78
ampullae, 5, 71, 72, 74
ampulla of Lorenzini, 75, 78
anguilliform, 1
anterior cardinal sinus, 24, 50
anterior cerebral artery, 46
anterior chamber, 77, 78
antiarchiformes, vii
apertures, 5
arachnoid, 61
archinephric duct, 54, 56, 58, 59
arterial circle, 46
artery
    afferent branchial artery, 43–45
    anterior cerebral artery, 46
    arterial circle, 46
    basilar artery, 46
    branchial, 43–45
    carotid, 46
    caudal intestinal artery, 37, 48
    caudal mesenteric artery, 47, 49, 56
    celiac artery, 31, 45, 49
    coeliac, 47, 49
    conus arteriosus, 41, 43
    coronary artery, 37, 43, 46
    cranial intestinal artery, 37, 47
    duodenal, 39, 47
    efferent branchial arch, 45, 46
    efferent branchial artery, 43–45
    efferent renal artery, 56
    external carotid artery, 45
    gastric artery, 37
    gastrosplenic, 47
    hepatic artery, 47, 49
    hyoidean, 45, 46
    hyoidean epibranchial artery, 45
    hypobranchial artery, 43, 45
    iliac, 47, 49
    internal carotid, 46, 66
    intertrematic artery, 44

lienogastric, 47, 49
pancreaticomesenteric artery, 47, 49
posttrematic, 43–46
pretrematic, 43–46
renal artery, 53, 56
stapedial, 45, 46
subclavian artery, 43, 45, 47
ventral aorta, 43, 45
ventral carotid artery, 45
ventral gastric artery, 49
arthrodires, xi
arthrodiriformes, vii
atrial, 42
atrioventricular, 43
atrium, 41–43, 50
auditory, 10, 63–66, 73
auricle, 66, 68
aves, viii
axons, 78

basal plate, 7, 61
basapophysis, 14
basibranchial, 11, 12, 25, 41, 42
basihyal, 12, 25
basihyoid, 11
basilar artery, 46
basipophyses, 13
batoidea, ix
bile duct, 30, 31, 37, 39, 47, 52
bonnethead, 60
bowfin, 1
brachial, 19, 47, 51, 52
brachial vein, 51, 52
branchial adductor, 24, 34
branchial arches, 12, 20, 24, 32, 36, 40, 45
branchial artery, 43–45
buccal, 4, 64, 71
buccal canal, 4, 71
buccopharyngeal, 53

calcifications, 9
canthi, 5
canthus, 5
carangiform, 1
carina, 10
carotid, 9, 46
cartilage
    basapophysis, 14
    basibranchial, 11, 12, 25, 41, 42
    basihyal, 12, 25
    basihyoid, 11
    basipophyses, 13
    branchial arches, 12, 20, 24, 32, 36, 40, 45
    carina, 10
    centrum, 13, 14, 70
    ceratobranchial, 11, 24, 25, 32
    ceratobranchials, 12, 32
    ceratohyal, 12, 25, 36, 49
    ceratohyoid, 11
    chondrocranium, 9, 10, 12, 19, 46, 61, 67, 73, 74
    coracoarcual, 25
    coracobranchial, 25, 49
    coracoid, 11, 13, 15, 19–21, 25–28, 40–42
    epibranchial, 11, 12, 24, 32, 36, 44–46
    exoccipital, 24
    hyoid, 12, 18, 20, 21, 36, 52
    homandibula, 11, 12, 14, 22, 36
    hypobranchial, 11, 20, 25, 40, 43, 45, 46, 63, 65, 70
    interbranchial, 5, 34, 43
    intercalary, 13, 14, 69 70
    interorbital, 51, 52
    labial cartilage, 11
    mandibular, 4, 12, 20, 36, 63, 64
    meckel's cartilage, 11–13, 20–22, 25, 33, 43
    mesopterygium, 11, 13, 15
    metapterygia, 13
    metapterygium, 15, 26, 27
    nasal capsule, 9–11, 13
    neurocranium, 21, 22
    optic pedicel, 9
    palatoquadrate, 11–13, 20–22, 36
    palatoquadrati, 18, 20–23
    pharyngobranchial, 11, 12, 24
    propterygium, 11, 15
    pterygioquadrate, 12
    puboischadic bar, 26
    puboischiac bar, 15
    puboischiatic, 26
    quadratoarticular, 12
    rostrum, 5, 9–11, 13, 20, 22, 45, 61, 71, 74, 75
    scapula, 11, 13
    scapular, 13, 14, 19–21, 26
    scapulocoracoid, 11, 13, 21, 22, 25
    supraorbital, 4, 10, 71
    suprascapular, 13, 15
caudal, x, xi, 1–3, 5, 8, 10, 12–14, 16, 19–21, 25, 28, 30, 32, 37, 40, 41, 43, 45, 47–54, 56, 58–60, 67–72, 74, 77, 78
caudal cardinal sinus, 47, 51
caudal fin, 1–3, 5, 8
caudal intestinal artery, 37, 48
caudal intestinal vein, 37, 49, 52
caudal ligament, 54, 58
caudal mesenteric artery, 47, 49, 56
caudal vein, 50, 52, 53
celiac artery, 31, 45, 49
cenozoic, viii
central canal, 69, 70
centrum, 13, 14, 70
ceratobranchial 11, 24, 25, 32
ceratobranchials, 12, 32
ceratohyal, 12, 25, 36, 49
ceratohyoid, 11
ceratotrichia, 13, 15, 26, 27
cerebellum, 62, 66–69, 73
cerebral hemisphere, 62, 66, 68
cerebrospinal fluid, 61, 69
cerebrum, 46, 67, 68
chondrichthyes, vii, viii, ix, 30, 58
chondrocranium, 9, 10, 12, 19, 46, 61, 67, 73, 74
chondrocytes, 9
choroid plexus, 67–69
ciliary body, 78

clasper, 4–6, 13, 15, 48, 55, 60
claspers, 5, 13, 26, 60
cloaca, 5, 6, 16, 28, 39, 47, 54, 56, 58–60
cloacal, 2, 5, 47, 52
cochlea, 74
cochlear, 78
coeliac, 47, 49
coelom, 29, 30, 54
coelomic, 28–32
collecting tubule, 56
columnar, 7, 8, 38, 39
commissural, 46
condyle(s), 9, 10
conjunctiva, 5, 77
constrictor hyoideus, 22
conus arteriosus, 41, 43
coprodeum, 60
copulation, 5, 60
coracoarcual, 25
coracoarcuales, 21, 23
coracobranchial, 25, 49
coracobranchiales, 23
coracohyoideus, 23, 25
coracoid, 11, 13, 15, 19–21, 25–28, 40–42
coracomandibularis, 23, 25
cornea, 75, 77, 78
coronary artery, 37, 43, 46
cranial, x, xi, 1, 10, 14, 16, 19, 21, 24, 25, 28, 30, 32, 37, 39, 41, 47, 49, 51–54, 59–61, 63, 67, 70
cranial cardinal sinus, 51
cranial intestinal artery, 37, 47
cranial intestinal vein, 37
cranial mesenteric artery, 47, 49
cretaceous, viii
cristae, 65, 72, 74
cuboidal, 6, 7, 38
cucullaris, 18, 20–22, 24
cupula, 72, 74
cutaneous, 52
cystic, 39

dentin, 8, 32, 33, 35
dentine, 7, 32, 35
dentition, 8
dermatocranium, 9
dermis, 5–8, 12, 32
devonian, vii, viii
diastole, 42
diencephalon, 62, 64, 66–68
digitiform, 30
distalis, 66
dogfish, ix, 4–6, 14, 16, 27, 35, 36, 53, 57, 60, 70
dorsal, x, xi, 1–5, 9, 10, 12, 14, 16, 17, 19–21, 24–26, 28–31, 34, 39, 40, 41, 43, 45–47, 49–56, 59–61, 66–70, 72–75, 78
dorsal aorta, 43, 45–47, 49–51, 54, 55
dorsal constrictor, 20, 34
dorsal fin, 4, 70
dorsal horn, 70
dorsalis trunci, 16, 18
dorsal mesentery, 29–31, 47

dorsal root, 14, 69, 70
dorsal root foramen, 14
dorsal root ganglion, 69, 70
duct
    archinephric, 54, 56, 58, 59
    bile duct, 30, 31, 37, 39, 47, 52
    cystic, 39
    endolymphatic, 5, 9, 10, 72, 73
    hepatic duct, 39
    oviduct, 49, 54, 56, 58–60
    perilymphatic, 9, 10, 72, 73
ductules, 59
duodenum, 30, 36, 38, 39, 47, 52
durodentine, 7

ectomesenchymal, 35
efferent branchial arch, 45, 46
efferent branchial artery, 43–45
efferent renal artery, 56
ejaculation, 5
elasmobranch, xi, 3, 35, 53, 57, 60
enamel, 7, 32, 33
endochondral, 32
endocranium, 9
endoderm, 36
endolymph, 74
endolymphatic, 5, 9, 10, 72, 73
endolymphatic duct, 73
endolymphatic fossa, 9, 10
endoskeletal, 14
epaxial, 16, 18, 19, 21, 24, 71, 73
epibranchial, 11, 12, 24, 32, 36, 44–46
epidermal, 6–8
epidermis, 5–8, 77
epididymis, 59
epiphyseal, 9, 10
epithelium, 7, 35, 36, 38–40, 59, 60, 67, 74, 75
excurrent aperture, 4, 5
exoccipital, 24
external carotid artery, 45
exudate, 5
eye, 4, 5, 9, 10, 20, 51, 61, 66, 67, 71, 74, 77, 78
eyelid, 4

facial, 10, 20, 63, 64
falciform, 28, 30–32, 37, 58, 59
fascia, 20, 21, 26
fenestrae, 10
fibroblasts, 39
foramen, 9–11, 14, 24, 46, 68, 69
    carotid, 9
    dorsal root foramen, 14
    epiphyseal, 9, 10
    foramen magnum, 9, 10, 24
    glossopharyngeal, 9, 10, 63, 65, 68
    occipital, 9, 10, 21, 63, 70
    ophthalmic, 9, 10, 64, 66
    optic foramen, 10, 11
    perilymphatic, 9, 10, 72, 73
    vagus, 9
    ventral root foramen, 14
foramen magnum, 9, 10, 24
forebrain, 68

fossa, 9, 10, 69
funculus, 70
fundic, 31, 36–38, 47, 49
funiculi, 69, 70

gallbladder, 31, 37, 39, 49
gastric artery, 37
gastric vein, 37, 49, 52
gastrohepatic ligament, 30, 31, 47
gastrohepatoduodenal, 52
gastrosplenic, 47, 52
germinative, 6, 7
gill lamellae, 5, 12, 34, 43
gill raker, 34
gill ray, 34
gill slits, 4, 5, 12, 19, 32, 40, 46
gland
    adenohypophysis, 66, 67
    nidamental, 59
    pars distalis, 66
    pars nervosa, 66, 67
    pituitary, 63, 66–69
    rectal gland, 37, 47–49, 52, 54, 56, 57
glomerulus, 53, 54, 56
glossopharyngeal, 9, 10, 63, 65, 68
gnathostomata, ix
gnathostomes, 12
gonadal vein, 51

habenula, 68
habenular nucleus, 69
haemal arch, 13, 14, 52
hemibranchs, 32
hepar, 39
hepatic, 30, 39–41, 47, 49–52
hepatic artery, 47, 49
hepatic duct, 39
hepatic portal vein, 30, 49, 52
hepatic sinus, 41, 50, 51
hepatic vein, 40, 51
hepatoduodenal, 30, 31, 39
heterocercal, 1–4
holobranch, 32
holonephric, 54
holonephros, 54
homocercal, 2, 3
humour, 78
hydroxyapatite, 9
hyoid, 12, 18, 20, 21, 36, 52
hyoid constrictor, 18, 20, 21
hyoidean, 45, 46, 52
hyoidean epibranchial artery, 45
hyomandibula, 11, 12, 14, 22, 36
hyomandibular, 12, 21, 36, 63, 64
hyomandibular canal, 4, 71
hypaxial, 16, 18, 19, 71
hypobranchial, 11, 20, 25, 40, 43, 45, 46, 63, 65, 70
hypobranchial artery, 43, 45
hypobranchial nerve, 63
hypocercal, 3, 4
hypochordal, 3, 4
hypoglossal, 65

iliac, 26, 47, 49, 50, 52, 53
iliac artery, 49
iliac vein, 50
incurrent aperture, 4, 5
inferior jugular vein, 51, 52
inferior oblique, 63, 64, 75–77
inferior rectus, 63, 75–77
infraorbital canal, 71
infraorbital trunk, 63
infundibulum, 63, 66, 68
interarcual, 24
interarcuales, 23
interarcualis dorsalis, 24
interbranchial, 5, 34, 43
interbranchial muscle, 34
intercalary, 13, 14, 69, 70
intercalary plate, 14, 70
interhyoideus, 21–23, 25
interlamellar, 45
intermandibularis, 20–23, 25
internal carotid, 46, 66
interorbital, 51, 52
interrenal, 57
intertrematic artery, 44
interventricular, 68
intraarcuals, 24
intraorbital canal, 4
invagination, 7
iris, 75, 77

jurassic, viii

keratinized, 6
kinocilium, 72

labial cartilage, 11
labyrinths, 72
lagena, 74
lamellae, 5, 10, 12, 32, 34, 43
lamellar, 32, 45
lamina propria, 38, 39
lateral abdominal vein, 47–52, 55
lateralis, 5, 67, 71, 78
lateral line, 4, 5, 16, 51, 71, 72, 78
lateral rectus, 66, 75, 76
lens, 77
levator hyomandibulae, 18, 23
levator hyomandibuli, 21, 22
levator palatoquadrati, 18, 20–23
lienogastric, 30, 47, 49, 52
lienogastric artery, 49
lienogastric vein, 49
linea alba, 17, 21, 26, 27
liver, 28, 30, 31, 39, 40, 47, 48, 51, 52, 58, 59
lorenzini canals, 64
lumbrosacral, 70

macula, 74
maculae, 65, 72, 74
macular, 74
mammalia, viii
mandibular, 4, 12, 20, 36, 63, 64
mandibular canal, 4

Meckel's cartilage, 11–13, 20–22, 25, 33, 43
median rectus, 76
medulla oblongata, 46, 62, 63, 66–69, 73
mesencephalon, 61, 62, 64, 67, 68
mesenchymal, 5, 7
mesenteries, 28–30, 32, 40, 59
mesentery, 29–31, 36, 47, 59
    dorsal mesentery, 29–31, 47
    falciform ligament, 28, 30–32, 37, 58
    gastrohepatic, 30, 31, 47
    gastrohepatic ligament, 30, 31, 47
    hepatoduodenal, 30, 31, 39
    mesoduodenum, 36
    mesogaster, 30, 31, 36
    mesointestine, 30
    mesorchium, 30, 55, 59
    mesorectum, 30, 47, 56
    mesotubarium, 30, 58, 59
    mesotubarium, 30, 58, 59
    mesovarium, 30, 31, 58, 59
    omentum, 52
    pericardium, 30, 40
    ventral mesentery, 29–31
mesoderm, 36
mesodermal, 7
mesoduodenum, 36
mesogaster, 30, 31, 36
mesointestine, 30
mesonephric, 54
mesonephros, 54
mesopterygium, 11, 13, 15
mesorchium, 30, 55, 59
mesorectum, 30, 47, 56
mesotubarium, 30, 58, 59
mesovarium, 30, 31, 58, 59
metanephros, 54
metapterygia, 13
mctapterygium, 15, 26, 27
metencephalon, 62
mississippian, viii
mouth, 4, 5, 9, 10, 12, 20, 32, 36, 40, 41, 45, 51, 61, 66
mucosal, 38, 40
mucus, 6, 7, 39, 74
muscle
    adductor branchiae, 23, 24
    adductor mandibulae, 18, 20–23
    adductor muscles, 70
    branchial adductor, 24, 34
    constrictor hyoideus, 22
    coracoarcuales, 21, 23
    coracobranchiales, 23
    coracohyoideus, 23, 25
    coracomandibularis, 23, 25
    cucullaris, 18, 20–22, 24
    dorsal constrictor, 20, 34
    dorsalis trunci, 16, 18
    hyoid constrictor, 18, 20, 21
    inferior oblique, 63, 64, 75–77
    inferior rectus, 63, 75–77
    interarcual, 24

interarcuales, 23
interarcualis dorsals, 24
interbranchial muscle, 34
interhyoideus, 21–23, 25
intermandibularis, 20–23, 25
lateral rectus, 66, 75, 76
levator hyomandibulae, 18, 23
levator hyomandibuli, 21, 22
levator palatoquadrati, 18, 20–22
median rectus, 76
myocommata, 16
myomere, 19, 27
myotome, 16–19, 21, 51
myotomes, 16–19, 47, 52
pectoral depressor, 21, 26, 27
pectoral levator, 18, 26, 27
pectoral levators, 20
pelvic adductor, 26, 27
pelvic depressor, 26, 27
pelvic levator, 26, 27
preorbital, 20, 36
preorbitalis, 20–23
protractor lentis, 64, 77, 78
rectus abdominis, 17
septal constrictor, 34, 44
septal constrictors, 20
suborbitalis, 22
subspinalis, 24
subvertebral, 17, 18
superior oblique, 64, 66, 75–77
superior rectus, 66, 75–77
ventral constrictor, 20, 34
ventral constrictors, 18, 20
ventral hyoid constrictor, 21
muscularis externa, 38, 39
muscularis mucosae, 38, 39
myelencephalon, 62
myelinated, 69
myocommata, 16
myomere, 19, 27
myosepta, 16, 19, 27, 47
myoseptum, 18, 19
myotome, 16–19, 21, 51
myotomes, 16–19, 47, 52

nares, 5, 10
nasal capsule, 9–11, 13
nasal flap, 5, 75
nephrostomes, 54
nerve
    abducens, 63, 64, 68, 69, 76
    auditory, 10, 63–66, 73
    buccal, 64
    dorsal root, 14, 69, 70
    facial, 10, 20, 63, 64
    glossopharyngeal, 9, 10, 63, 65, 68
    hyomandibular, 4, 12, 21, 36, 63, 64, 71
    hypobranchial, 63, 65
    hypoglossal, 65
    infraobital trunk, 63
    mandibular, 4, 12, 20, 36, 63, 64
    nervus terminalis, 64
    occipital nerves, 63
    oculomotor, 63, 64, 68
    olfactory, 10, 46, 62–64, 66–69, 71, 74, 75
    olfactory tract, 63, 66–69, 74
    ophthalmic, 9, 10, 64, 66
    optic, 9–11, 46, 62–64, 66–69, 75, 77, 78
    optic nerve, 46, 63, 66, 68, 69, 75, 77, 78
    palatine, 64
    spinal nerves, 19, 63, 69, 70
    splanchnic, 36
    temporomandibular, 14
    trigeminal, 63, 64
    trochlear, 64, 66–68, 75, 76
    vagus, 9, 10, 63, 65, 68, 70
    ventral root, 14, 69, 70
nervus terminalis, 64
neural arch, 13, 14
neural canal, 14, 61, 69
neural spine, 13, 14
neurocranium, 21, 22
neuromast, 72
neuromasts, 72
neurons, 69, 74
nidamental, 59
nostril, 4
notochord, 13, 14

occipital, 9, 10, 21, 63, 70
occipital condyle, 9, 10
occipital nerves, 63
oculomotor, 63, 64, 68
odontoblasts, 33
olfactory, 10, 46, 62–64, 66–69, 71, 74, 75
olfactory epithelium, 75
olfactory lobe, 62, 66–69
olfactory sac, 66, 74, 75
olfactory tract, 63, 66–69, 74
omentum, 52
ophthalmic, 9, 10, 64, 66
opisthonephric, 54, 59
opisthonephros, 54, 59
optic, 9–11, 46, 62–64, 66–69, 75, 77, 78
optic chiasma, 63, 66, 67, 69
optic foramen, 10, 11
optic lobe, 62, 66–69
optic nerve, 46, 63, 66, 68, 69, 75, 77, 78
optic tectum, 69
orbital process, 9–12, 61
ordovician, viii
osteichthyes, viii, ix
osteocytes, 9
ovary, 30, 58, 59
oviduct, 49, 54, 56, 58–60

palatine, 64
palatoquadrate, 11–13, 20–22, 36
palatoquadrati, 18, 20–23
pancreas, 31, 37, 39, 49, 52
pancreaticomesenteric, 30, 47, 49, 52

pancreaticomesenteric artery, 47, 49
pancreaticomesenteric vein, 49
papillae, 8, 32, 36, 56
paraphysis, 68
parapophyses, 13
parietal peritoneum, 29
parietal vein, 51
pars distalis, 66
pars nervosa, 66, 67
pectoral depressor, 21, 26, 27
pectoral fin, 4, 5, 13, 16, 47, 52, 70
pectoral girdle, 13
pectoral levator, 18, 26, 27
pectoral levators, 20
pedicel, 9, 10, 74–77
peduncle, 67
pelvic adductor, 26, 27
pelvic depressor, 26, 27
pelvic fin, 4, 5, 13, 52, 60, 70
pelvic girdle, 13, 28, 52
pelvic levator, 26, 27
pericardial cavity, 30, 40–43
pericardium, 30, 40
perichondrium, 9
perilymph, 74
perilymphatic, 9, 10, 72, 73
peritoneal, 30, 40, 54
peritoneum, 29, 36, 59
permian, viii
pharyngobranchial, 11, 12, 24
pia mater, 61
pineal, 10, 68
pituitary, 62, 66–69
placental, 58
placentation, 60
placodermi, vii, ix
placoid, 5–7, 32
pleuroperitoneal, 30, 39, 40, 47, 52, 54, 59
pleuroperitoneum, 39
postcardinal sinus, 50
postcardinal vein, 50
posterior chamber, 77, 78
postganglionic, 56
postotic process, 9
posttrematic, 43–46
precerebral cavity, 10, 75
preorbital, 20, 36
preorbitalis, 20–23
pretrematic, 43–46
process
    adductor, 11
    orbital, 9–12, 61
    postotic, 9
    scapular, 13, 14, 20, 26
proctodeum, 56, 58, 60
pronephric, 54, 57
pronephros, 54
propterygium, 11, 15
prosencephalon, 61, 62
protractor lentis, 64, 77, 78
pseudobranch, 46
pseudobranchs, 35

pterygiophores, 13, 26
pterygioquadrate, 12
puboischiadic bar, 26
puboischiac bar, 15
puboischiatic, 26
pupil, 75, 77, 78
pyloric, 31, 36–38, 47, 49, 52
pylorus, 36, 52

quadratoarticular, 12

radials, 13, 15
rakers, 35
raphe, 20, 21
rectal gland, 37, 47–49, 52, 54, 56, 57
rectum, 30, 37, 39, 56, 58
rectus abdominis, 17
renal artery, 53, 56
renal corpuscle, 53, 56
renal portal, 50, 52–54, 56
renal tubule, 56, 57
renal vein, 56
reptilia, viii
restiform, 67
rete mirable, 16
retina, 66, 74, 77, 78
retroperitoneal, 54
rhombencephalon, 62, 68
rhomboidea, 69
rostral, x, xi, 5, 10, 12, 16, 20, 21, 25, 30, 41, 43, 45, 46, 54, 67, 68, 71, 74, 77
rostrum, 5, 9–11, 13, 20, 22, 45, 61, 71, 74, 75
rugae, 36, 38

saccule, 72
sacculus, 73, 74
saccus, 63, 66–69
saccus vasculosus, 63, 68
sagittal, x, xi, 42, 58, 68, 69, 77
scapula, 11, 13
scapular, 13, 14, 19–21, 26
sacpular process, 13, 14, 20, 26
scapulocoracoid, 11, 13, 21, 22, 25
sclera, 77, 78
semicircular canal, 73, 74
seminal vesicle, 48, 55, 56, 59
septal constrictor(s), 20, 34, 44
serosa, 38
serotonin, 60
shell gland, 58, 59
silurian, viii
sinoatrial, 43
sinus venosus, 40–43, 49, 51
siphon chamber, 48
siphon sac, 5, 6, 55, 60
spermatic, 47, 60
spermatogenesis, 60
spermatozoa, 59
spermatozoon, 60
spermiogenesis, 60
sperm sac, 55
spinal nerves, 19, 63, 69, 70

spine, 1, 4–7, 13, 14
spines, 1, 5, 6, 16
spiracle(s), 4, 5, 16, 19–21, 32, 46, 51, 71
spiracular, 5, 20, 21, 32, 35, 46
spiral intestine, 28, 30, 31, 37, 39, 49
splanchnic, 36
spleen, 30, 31, 37, 39, 47, 49, 52
squalus, ix, 1, 2, 8, 12, 35, 36, 53, 57, 60
stapedial artery, 45, 46
stereocilia, 72
stomach, 28, 30–32, 36–39, 41, 47–49, 52, 53
subclavian, 43, 45, 47, 50–52
subclavian artery, 43, 45, 47
subclavian vein, 50–52
subdural, 69
suborbitalis, 22
subscapular vein, 51
subspinalis, 24
subvertebral, 17, 18
sulcus, 60
superior oblique, 64, 66, 75–77
superior rectus, 66, 75–77
supraorbital, 4, 10, 71
supraorbital canal, 4, 71
suprascapular, 13, 15
suprascapular cartilage, 13, 15
suspensory ligament, 77, 78
synaptic, 69
systole, 42

tectum, 67, 69
tegmen, 9, 10
tegmen cranii, 9, 10
tela choroidea, 67–69
telencephalon, 62, 63, 66–68
teleost, 2, 3, 35
temporomandibular, 14
tesserae, 9, 14, 33
transverse septum, 13, 16, 18, 19, 41, 47
triassic, viii
trigeminal, 63, 64
trimethylamine oxide, 40, 54, 57
trochlear, 64, 66–68, 75, 76
tunica mucosae, 38
typhlosole, 38, 39

urinary papilla, 6, 54, 56, 58, 59
urodeum, 56, 58, 60
urogenital, 5, 6, 55, 56, 59, 60
urogenital papilla, 5, 6, 55, 56, 59, 60
uteri, 30, 56, 59
uterus, 28, 31, 58–60
utricule, 72
utriculus, 72–74

vagus, 9, 10, 63, 65, 68, 70
valvular, 36, 39, 47, 52
vas deferens, 48, 55, 56, 59
vein
    anterior cardinal sinus, 24, 50
    brachial vein, 51, 52
    caudal cardinal sinus, 47, 51

caudal intestinal vein, 37, 49, 52
caudal vein, 50, 52, 53
cranial cardinal sinus, 51
cranial intestinal vein, 37
gastric vein, 37, 49, 52
gonadal vein, 51
hepatic portal vein, 30, 49, 52
hepatic sinus, 41, 50, 51
hepatic vein, 40, 51
iliac vein, 50
inferior jugular vein, 51, 52
lateral abdominal vein, 47–52, 55
lienogastric vein, 49
pancreaticomesenteric vein, 49
parietal vein, 51
postcardinal, 50

postcardinal sinus, 50
renal portal, 50, 52–54, 56
renal portal vein, 50, 53, 56
renal vein, 56
sinus venosus, 40–43, 49, 51
subclavian vein, 50–52
subscapular vein, 51
ventral gastric vein, 49
ventral, x, xi, 1–6, 9–21, 23–26, 28–31, 34, 37–41, 43, 45–52, 55, 56, 58, 59, 61, 63, 66–71, 74, 75, 78
ventral aorta, 43, 45
ventral carotid artery, 45
ventral constrictor, 20, 34
ventral constrictors, 18, 20
ventral gastric artery, 49

ventral gastric vein, 49
ventral horn, 70
ventral hyoid constrictor, 21
ventral mesentery, 29–31
ventral root, 14, 69, 70
ventral root foramen, 14
ventricle, 41, 43, 50, 66–69
vesicles, 56, 59
villi, 39, 59
visceral peritoneum, 29
vitreous chamber, 77, 78
vitrodentine, 7
viviparity, 58
viviparous, 58

zonule fibers, 77